ZENITH

The Science, History and Mythology of the Northern Lights

Liam Devonport

Praise for *Zenith*

#1 New Release on Amazon

(*Arctic and Antarctic History* and *History of Astronomy*)

'An astounding tale that showcases humankind's desire to cure curiosity and understand the heavens.'

Glossop Chronicle

'A must-read for anybody eager to learn more about this astonishing wonder of the world.'

Around Saddleworth and Tameside Magazine

'Fantastic reading. Insightful and thought provoking. Written with a great sense of humour.'

Joe, Amazon review

'This is a fun yet informative piece of work. Easy to read, excellently put together, full of interesting facts I would never have found online and sprinkled with humour. In short, it was a delight to read.'

Stephen, Amazon review

'Move over Bill Bryson, there's a new kid in town'.

David, Amazon review

Copyright © 2020 Liam Devonport

All rights reserved.

No part of this book may be reproduced, or stored in a retrieval system, or transmitted in any form or by any means, electronic, mechanical, photocopying, recording, or otherwise, without express written permission of the publisher.

First published in Great Britain

All images used are royalty free and sourced with thanks from Pixabay.

A special thanks is given to all contributing photographers:

Neale LaSalle
Rudolf Kirchner
Stein Egil Liland
Tobias Bjørkli
Visit Greenland
Wikilmages

Liam Devonport asserts the moral right to be identified as the author of this work.

To my wife, Roxana – Hubco

Contents

Preface	1
A whistle stop tour of history	6
Myths and tales	16
The beginning of the end	32
The curious life of Kristian Olaf Bernhard Birkeland	44
Standing on the shoulders of giants	70
The science bit	75
The atmosphere	84
The great geomagnetic storm	93
Beyond earth	102
Photographs	106
Acknowledgements	120
Connect with the author	122
Select bibliography	124

'Some believe that whistling and making other sounds at the aurora will either cause it to become more active, use it as a way to speak to their ancestors or even that the aurora will come down and take their heads off, thus making them observe it in silence and awe.'

Andy Long, Photographing the Aurora Borealis: How to Shoot the Northern Lights

PREFACE

Now you've just picked up this book, I want you to put it down and turn on the closest lamp to you. Go on, put it down. It's only the preface so we haven't started yet.

The bulb you have just turned on is likely somewhere between 40-60 watts. So, chances are it will look reasonably bright, especially if you're looking at a naked bulb as opposed to one encased by a shade. Don't worry if you're not as I'm not going to ask you to stand up again. Just take note of roughly how bright the bulb is. Now, I want you to take that bulb and multiply its brightness by twenty million, which will give you an astounding one billion watts of energy. Whilst you're trying to do this, I think it would be safe to say that such an amount of energy will be hard to visualise, as the numbers are so humongous, they surpass your mind's limit of visual comprehension. At one billion watts, the energy emitted would be such that you would have to vacate the bulb's immediate surroundings and seek refuge in whatever shade remained. This would be easier said than done, because one billion watts of energy would be sufficiently prolific to be seen from the beginnings of space, 100km above Earth's surface. Now, if you were to take said bulb, launch it into space, and once it hit the outer regions of the atmosphere you were to smash it to pieces releasing all one billion watts of energy, the results would be akin to that of the aurora borealis. Not only that, but this bulb-induced aurora would be visible to many people around the globe from Scandinavia to North America and beyond. Thousands would stare into the sky, both amazed and seduced by the hypnotic allure of the aurora. And whilst they stood with necks craned upwards, witnessing every flick, flash, and fold of the lights, memorizing every colour, movement, and shape they made, their bodies would be subconsciously reacting to the energy of the lights in ways they would find a little surprising, if not a little disturbing. You see, every one of us is affected by the aurora in a surprisingly similar way. And I'm not just referring to the science, history, and mythology of them, which we will get to in time. I'm talking about the physiological influence they have on you as an organic being. You see, when you think about it, as a fleshy bag of bones (no offense), which is influenced by its environment, it should not come as a surprise to learn that solar activity and the auroras can influence you. That's right, I'm talking to *you*. As one billion watts of energy cuts its way through the night sky leaving dancing trails of luminosity in its wake, some of the radiation and high-energy particles will inevitably make their way towards Earth's surface, and in turn, into you. I would like to state at this point that this is probably nothing to worry about. Earth's magnetic field does a rather splendid job of protecting you from solar radiation by acting as a giant filtration system that shields you from the ills of the Universe.

Let's say you're an eager aurora enthusiast and you've heard there's a particularly spectacular display on its way. You make your way to the Arctic circle, climb to the highest point you can find, face north into a bracing wind, and wait. The sky is lucidly clear, the tundra deafeningly quiet, and the Universe above an endless depth of impenetrable black. Much to your relief you don't have to wait long before the aurora starts to appear. And once the early glow of their arrival begins to creep on the northern horizon, you immediately submit to their alluring spell. They start dimly at first. Nothing more than a faint green haze low in the sky. But before long they begin to grow brighter and more powerful, until they consume the night sky in a hive of rhythmic activity. Your mind starts to wonder as you enter a hypnotic state of bliss. The lights are stunning, possibly one of the most beautiful things you've ever seen, and as you stand alone in the tundra bracing the wind, but happy in the tranquility and stillness of the moment, you literally *feel* them coursing through your body. The lights spare no effort performing their well-rehearsed routine of extraterrestrial divinity, which is when you realise, much to your delight, you're entirely alone, and this unique performance was entirely for you.

As your mind wanders your imagination begins to race, whilst all the while, your soul achingly stirs at the beauty of the lights. Your body is also feeling the profound effects of this intense solar energy. But the physical sensation you're experiencing is not just being instigated by your mind. Not at all. It's more than just a figment of your imagination. The connection between solar energy and the human body is a well-known affair, which was documented as far back as 1926 by the Russian Professor, Alexander Tchijevsky, when he presented his findings to the American Metrological Society at their annual meeting. In the paper he presented named *Physical Factors of the Historic Process*, Tchijevsky wrote:

> *The Sun is an enormous generator of electric energy and emits it in the form of radiation and induction. The Sun is surrounded by an electromagnetic field, the limits of which reach beyond the farthermost planet Neptune, and therefore the Earth with its electromagnetic field is in the Sun's field of tremendous power.*

It's an obvious, welcoming, and exciting prospect (if not slightly unnerving) that we are within such proximity to the Sun and its energy. As you can imagine, being this close to the Sun means we are in the firing line of its activity, whether that happens to be placid or belligerent. Both Earth and humanity are at the whim of the Sun for better or worse. But this fact brings us to a humbling conclusion: when the Sun releases large amounts of solar energy, both Earth and humans react as one. What affects Earth affects us, and vice versa, if only to a significantly lesser degree. We are intricately connected to this giant spinning rock we call home, and that's not about to change in a hurry. Think of it this way: the aurora is one of Earth's reactions to aggressive solar activity; albeit a beautiful, rousing reaction. And as we have just learnt, as the Earth and we are connected, as Earth reacts, so do we. Albeit not quite as spectacularly. Tchijevsky continues to say that solar activity plays a role 'in the chaining of the heartbeat, in the alterations of the chemical composition of the blood...' and that these activities 'result in the general condition of the organism [human body] and its nervous toxicity'. This means that as the atmosphere varies in thickness depending on where you are on Earth, and the magnetic field varies in strength at different locations across the globe, where you happen to reside can influence how affected you are by solar activity. And the more powerful the aurora, the greater you are affected physiologically.

NASA state that solar activity doesn't just produce the auroras, it also contributes to other atmospheric commotion such as lightning intensity, which can further affect your body. In an astounding piece of research, NASA stated that at any moment '2,000 thunderstorms roll over Earth, producing 50 flashes of lightning every second.' It's atmospheric volatility such as this that can affect your brain, heart, and central nervous system. Dr Neil Cherry reported in a paper in 2001 that some of the side effects of extreme solar activity can be 'altered blood pressure and melatonin...' as well as '...reproductive, cardiac and neurological disease and death.' But on a slightly less concerning note, milder symptoms such as increases in anxiety, sleep disturbances, changed moods, and more accidents have also been shown to happen on days with high solar activity and greater aurora movement. There's even an increase in crime and traffic incidents. *It's astonishing, isn't it?*

This realisation that the Universe and solar system's activity affects both Earth and us allows us to grasp an even more poignant thought: we are not just connected to Earth, we are connected to the entire Universe and everything within. From you to me, and from the Sun and Earth to beyond, we are all intricately linked in some way. And as we react to solar activity and the auroras, we in turn cause other things to react to us. We are all part of one giant, infinite chain of actions and reactions with an undecipherable beginning and an unpredictable end. It's a beautiful thought. A thought that has been articulated by many over the years, but by none greater than Lawrence M. Krauss, who solidifies this point both poetically and scientifically:

> *The amazing thing is that every atom in your body came from a star that exploded. And the atoms in your left hand probably came from a different star than your right hand. It really is the most poetic thing I know about physics: You are all stardust. You couldn't be here if stars hadn't exploded because the elements - the carbon, nitrogen, oxygen, iron, all the things that matter for evolution - weren't created at the beginning of time. They were created in the nuclear furnaces of stars, and the only way they could get into your body is if those stars were kind enough to explode. So, forget Jesus. The stars died so that you could be here today.*

A WHISTLE STOP TOUR OF HISTORY

'If you don't know your history, then you don't know anything. You are a leaf that doesn't know it is part of a tree.'

Michael Crichton

One thousand two hundred generations ago around the year 28,000BC an inquisitive, robustly built caveman with an artistic flare, in what is now known as south-west France, eagerly watched the sky at night. In modern scientific literature the cave people of this time are known as European early modern humans, and they are characterised by their powerful physique, large brain, straight forehead, and short, wide face. It was an incredibly important time to be alive in terms of early Homo sapiens. And although these early humans changed the course of humanity forever, it would be safe to say they were not renowned for their good looks. This period sits right in the centre of an age known as the Upper Paleolithic, or Late Stone Age, which is recognised as the era when humans adopted behavioral modernity. This was the moment in history when humans developed behavioral and cognitive traits that differentiated them from their primate lineage, such as music, art, abstract thinking, and depth of thought. It was the evolutionary time when we crossed the line from being a mildly intelligent biped to become the inquisitive, pondering, sarcastic, grumpy, innovative creatures we are today. A momentous moment in history – unless you ask my wife. Although at this time we were still predominantly a hunter-gatherer population, our newly evolved abilities allowed us to greater analyse our surroundings, contemplate meaning, and engage with the planet on a level previously uncharted. It was the moment that thrust humanity into the limelight and gave us the tools to eventually become Earth's apex species. As would be expected, however, all those years ago whilst our evolutionary advantages came to fruition, Earth would have been an entirely different place than it is today. Given the human population of this time was anticipated to have been less than one million, our species would have generated almost no light pollution, bar the odd little campfire here and there. This would have left the skies open, and the depths of the Universe exposed to the naked eye each night. It was this combination of clear skies and our new-found aptitude for profound thought that led to the now famous cave paintings of Cro-Magnon.

At some point around this period in history the primitive person in question would embrace their newly evolved intelligence and analytically watch the night sky. The black sheet of night that encased Earth would have been riddled with stars so clear and bright, we may struggle to imagine their sheer lucidity and abundance today. As they stared upward towards the heavens it's not possible to say what they were thinking, but one thing is for sure: at some point on one of those nights a series of colourful lights appeared as if from nowhere and filled the sky with inspiring luminosity. They shone with such purpose that the person below watched them dance, mentally recorded their movements, picked up a rock and possibly some charcoal, walked to a nearby cave, and began to draw. From this burning memory of the lights that filled both their mind and the night, their drawing continued until the walls of the cave progressively became one of the world's first works of art. But without doubt, and more importantly in the context of this book, the world's first documented recording of the aurora borealis.

It's still possible to see the painting today within the Cro-Magnon caves situated close to Dordogne, France, where the walls of these historically important caves are rightfully preserved for future generations. But even though these drawings are believed to be the first ever markings of the aurora, the deciphering of the painting does require a little imagination. As to the untrained eye the painting looks a little like the first attempt at drawing Mr Tickle. Regardless of this caveman's penchant for Mr Men, the documented obsession with the auroras had begun, and it was all thanks to one person with a rock, an imagination, and access to a cave which they envisaged as a canvas.

As we skip through the pages of history from this point forth, we realise that the documenting of the auroras becomes somewhat non-existent. Just as it had before our beloved caveman decided to draw them one night. And as we continue to look for documentation and evidence of their presence, we find the next observations (and possibly the first ever written documentation of the auroras) arose around 4,600 years ago. Albeit a briefly written passage, history, on this occasion, provides us with an incredibly short story. It is said that around four millennia ago the great landmass of China was protected by the guardian-like figure of the Emperor with four faces, otherwise known as the Yellow Emperor. The Yellow Emperor was a cosmological deity who was born from the skies of Earth, whose power and prowess were beyond the comprehension of mere mortals. He was known for his duty as the god of light and thunder: two of what were considered the most powerful aspects of nature. With his flashy dressing gown, well-oiled beard, robust staff, and fancy square party hat, few dared to cross such a mighty god's path. Yet the tale of his birth and his mother's impregnation is what takes this story in the direction of the auroras. This story, which is only a sentence in length, says that 'Fu-Pao, the mother of the Yellow Empire Shuan-Yuan, saw strong lightning moving around the star Su, which belongs to the constellation of Bei-Dou, and the light illuminated the whole area.' The lightning in this instance is considered to be the glowing of the auroras, which were weaving amongst one of the most visible constellations in the Chinese sky of that period. Once Fu-Pao had watched the auroras thread throughout the stars it was written that soon after '...she became pregnant'. This mysterious pregnancy, which was induced by the auroras, was believed to have been the body of the Yellow Emperor, who was later born into his rightful position as the god of light. In a rather poetic notion, as most Chinese mythology appears to contain, the god of light came *from* the light and was born *of* the light. And for thousands of years thereon, China was considered guarded by the son of the Northern Lights.

After Fu-Pao's pregnancy story, which sounds like the type of lie you could use to pull the wool over your parents' eyes before DNA tests were invented, the documenting of the auroras slips into a lull once again. And as the next 2,000 years soar by, we find nothing applicable to our story. Until all of a sudden, a poetic text appears:

> *As I looked, behold, a stormy wind came out of the North and a great cloud with brightness round about it, and fire flashing forth continuously and in the midst of the fire, as it were gleaming bronze like the appearance of the [rain]bow that is in the cloud on the day of rain, so was the appearance of the brightness round about.*

The author of this quote is reputed to have been a prophet-priest from Israel named, Ezekiel. Although this quote (or a translated variable of it) can be found within the Bible, it's believed by many academics to have been the first attempted, genuine, written documentation of the aurora borealis. It was written in the 6th century BC, six hundred years before the birth of Christianity, but was later taken and used in the formation of the Old Testament. It is further believed by some academics that there are in-fact up to five separate descriptions of the Northern Lights contained within the Bible, but this is a contested point.

This period of the above quote that ran from 600BC to 501BC, is considered to have been a pivotal age of thought for humankind. Buddah and Mahavira in India, Zoroaster in Persia, Pythagoras in Greece, and Confucius in China, formed what is commonly known as the 'five major thought streams', and these streams have more influence on our everyday lives than we would be inclined to believe. It's no surprise then that some of the first official written documentation of the auroras derives from this period; and that the citations of this time appear more philosophical, poetic, and thought provoking than a drawing of Mr Tickle. It's also important to note that now 27,400 years had passed since the first recording of the lights, the people who saw them now were no longer just looking at the auroras, they were analysing them.

It took roughly three centuries until any further notes were made on the lights, long after the death of Ezekiel. But this next brief noting on the lights by the ancient Greek philosopher, Plutarch, appears to be the tangible starting point of the true speculative documentation of the auroras. From this point forth, a succession of writings, documents, and beliefs have been found. Plutarch's writings appear to show a man who was sharp of mind, analytical in thought, and poetic in nature. His efforts to document the lights, in my opinion, are the truest and clearest we have seen on our journey so far. He wrote that:

> *During seventy days there was an enormous and furious figure in the sky. It was like a flaming cloud, which did not stay at its position but moved windingly and regularly, so that the glowing fragments were flying in all directions and fire was blazing as the comets do. Those fragments came loose during rushing and unexpected movements.*

Shortly after Plutarch's efforts, a further series of writings were made by another Greek philosopher who was born near Thessaloniki in 384BC. The child's father, Nicomachus, was a practicing physician who personally served King Amyntas of Macedon. But Nicomachus died whilst the boy was still a child, leaving him in the trustworthy hands of a family guardian. The child appeared to be a bright individual with an insatiable interest in academia, especially within the areas of science and philosophy. Once the boy had grown into a young man, he left his hometown and headed to Athens where he enrolled into Plato's Academy and studied for twenty years until age thirty-seven. You may not be surprised to learn at this point that the child's name was Aristotle. Aristotle witnessed the aurora in 344BC, and it was an occasion which evidently influenced him in a profoundly positive way, as he later wrote about the occurrence within his book *Meteorology*. Aristotle wrote that 'sometimes on a fine night we see a variety of appearances that form in the sky: 'chasms' for instance and 'trenches' and blood red colours. These, too, have the same cause. For we have seen that the upper air condenses into an inflammable condition and that the combustion sometimes takes on the appearance of a burning flame, sometimes that of moving torches and stars.' He also stated that these lights 'will cause all kinds of colours to appear, but especially crimson and purple. For these colours generally appear when fire-colour and white are combined by superposition.' 'These appearances do not last long' he writes, 'because the condensation of the air is transient.'

His observations and notes on the aurora were the most detailed and extensive ever made. His description of the lights is creditworthy, although he later makes a conclusion that is very far from the truth. He claims the auroras 'must be taken to be the cause of shooting stars'. Something he could not possibly have known was wrong at the time and would have therefore been a highly educated assumption. Aristotle also proclaimed that the heavens, being the Sun, Moon, and stars, never changed, insinuating that any natural occurrences, such as the auroras, could not have been caused by movements and fluctuations within the heavens and must be the result of alternations closer to home. This belief, in the words of the world-famous astronomer Carl Sagan (1934 – 1996), was the 'most influential error in the history of astronomy', as this presumed belief of heavenly stagnation confused natural philosophers for the coming 1,900 years. It essentially ruled out the possibility that space and its components had any effect upon Earth, which led to the eyes of science turning inwards towards our planet in a blinkered way. *And why not?* For at this point in history, Earth was considered to have been the centre of the Universe, and the idea that our planet was self-contained was the reigning dogmatic belief of the time. It wasn't until the year 1572 that the Danish astronomer, Tycho Brahe, witnessed the birth of a new star, which caused the minds of science to question what they thought they knew and turn their gaze outward. It is this willingness to change what is already culturally accepted that is the beauty of science. Its willingness to accept error and erroneous beliefs that has led to our greater understanding of the Northern Lights.

After Aristotle's documentation of the lights, it is not known if he ever witnessed them again, as no further description is offered and no additional investigation into shooting stars and their formation is mentioned. Besides, his simple theory that the auroras were created by evaporated water, which then collided with the fire element within the sky before combusting and creating the aurora was incorrect. But it's not hard to see why he formulated such a grounded opinion of the auroras, as opposed to attaching their materialisation to that of supreme beings, heavenly spirits, and Chinese gods in bathrobes. Aristotle was an avid academic, and during his years he studied a variety of subjects ranging from anatomy, astronomy, education, foreign customs, geography, geology, literature, meteorology, poetry, physics, and zoology. Meanwhile, within the realms of philosophy he wrote on aesthetics, ethics, government, metaphysics, politics, economics, psychology, rhetoric, and theology. Let's just say he was a very busy guy.

As well as the documentation left to us by artistic cavemen, the ancient Chinese, prophet-priests and Greek philosophers, the Northern Lights also left their mark upon history in a few other, beguiling ways. One such example occurred in 360BC when, Philip, the King of Macedon, was preparing a brutal attack on the city of Byzantium, which at the time was an ancient Greek colony, now Istanbul. The city of Byzantium was a significant base, encased by sturdy walls and stout defences. Gaining access to such a city was no easy feat. Not even for a King. But in a moment of creative enlightenment, King Philip decided the best way to infiltrate such a city would not be to destroy or clamber over its walls, but to dig under them. The dug tunnels would then be used as points of access for his men. Once the digging was complete, Philip's men flowed into the tunnels, crammed together like sardines, poised and ready to pounce at a moment's notice. But as midnight approached and the orders to strike drew closer, the night began to brighten in a rather unusual way. Out of the darkness a crescent moon-shaped light shone from above, illuminating the sky whilst casting shadows in its wake. The light flickered so brilliantly it allowed the citizens of Byzantium to sound the alarm as they noticed tunnels beneath the city's walls filled with anxious warriors. With the element of surprise distinguished, Philip knew his chances of success were squashed, so he led his army away in a stark retreat under the gleaming inconvenience of the aurora. Years later, after Philip's encounter with the Northern Lights, a special coin was forged to mark this serendipitous occasion, showcasing what is believed to have been an aurora arc. This arc, however, has been grossly misconstrued over the years and is now incorrectly believed to be symbolic of the Moon. This misinterpreted Moon was then taken and displayed proudly upon the Turkish flag alongside a five-pointed white star. Curiously, if this tale is to be believed, then somebody has a lot of explaining to do, as the Moon and star of the Turkish flag are considered highly representative symbols of Islam and are regularly used throughout the Islamic world. So, potentially, it turns out that the world's second largest religion could have nothing to do with the Moon and is visually represented by the aurora borealis instead. Now that's an awkward conversation waiting to happen.

As astonishing as it sounds, instances such as the above have been known to occur quite a few times throughout history. For example, in 44BC the Northern Lights shone so brightly as to deter a night attack from the Romans. It was reported that both cavalry and infantry soldiers were fully visible as the beams of light furiously illuminated the sky. It's also believed that when the Roman Emperor, Titus, destroyed Jerusalem, the lights burned as bright as flames in the skies of Palestine. Yet it's never been detailed as to whether the lights of this night were supposed to be symbolic of joy or sorrow. Even more dumbfounding than anything we have heard so far is that when the Roman Empire fell, the auroras disappeared, and Earth slipped into a thousand years of silence. A void of inactivity that is both known and recognised by science yet remains an enigma to this very day.

MYTHS AND TALES

'I believe that imagination is stronger than knowledge.'

Robert Fulghum

From this moment on as the years passed by, many people from a variety of nations would witness the auroras in action: Europeans, Americans, Australian aborigines, and Asians would all in some way or another note the passing of the lights. Whether it was via storytelling, music, drawing, or written documentation, records of the aurora's presence have been found in multiple locations around the planet. One of the most fruitful locations for astronomical text discovery is the old territory of Babylon – present day Hillah, roughly 50 miles south of Baghdad, Iraq. This ancient city appears to have nurtured a fascination for astronomy and marked its observations and discoveries on a series of baked (or Sun-dried) tablets in an almost forgotten cuneiform syllabic script, which is now only understood by a few hundred people on Earth. In total, it is estimated that between half a million and two million cuneiform tablets have been excavated, of which roughly 30,000 – 100,000 have been translated. It's a wonder what mysteries these yet to be translated tablets hold, and what observations they document. Evidently, if you speak the correct cuneiform script there's a lot of work out there for you.

But one of the chores of dealing with such age-old scripts as these is that all those years ago there appeared to be somewhat of an inclination for elaboration, fantasy, embellishment, and outright superstition. This means that any old documents relating to the Northern Lights must be taken with a considerably large pinch of salt. Nevertheless, these well salted tales prove superfluously entertaining although wholly inaccurate. But we shouldn't judge them too harshly, as these stories were the genuine beliefs that existed before the development of a solid foundation of science. And as such, they were the unwavering dogmas of their time.

One such theory that was attached to the aurora borealis was attributed to the Vikings and Norse mythology. The Viking age ran from roughly 800-1200AD, and throughout these years the Vikings would rule and conquer the lands of Scandinavia, parts of

Northern and Central Europe, and as far afield as European Russia, North Africa, and even the Middle East and Asia. Contrary to popular belief, however, the Vikings were not simply a savage race of blood-thirsty, bearded brutes. They were academics, inventors, and the forerunners of their time. They possessed highly developed seafaring skills, which enabled them to rule the seas and display their awesome power – all of which was characterised by their distinctive, fear inducing longships. These ships were purposely designed to cut through water with ease, enabling the Vikings to skim the surface of lakes and seas as quickly and effortlessly as possible. The ships were built from planks of oak, each overlapping and connecting to make a solid, robust body. These mighty vessels were further adorned with gracious wooden carvings of snakes or dragons, which once sculpted were mounted upon the ships' bows. Each ship was roughly 30 meters long and could carry up to 60 burly men. Large, square woven woolen sails were secured fast to the ships' masts, and each man was designated an ore to help propel the ship through water at speeds of up to ten miles per hour. Not much by today's standards, but fast and efficient for the time.

It's easy to see that the Viking's influenced and dominated many parts of British society during these times, because we still live with the aftereffects today, almost 1,000 years later. For example, many words contained within the English language such as axel, raft, leather, husband, knife, ransack, thorp and bylaw are derivative of Old Norse. So are a large collection of place names such as Egilsay, Ormskirk, Ravenscar, and Meols. The Vikings, their language, and mythology have helped shape the modern world in which we live. But more to the point, so have their beliefs and opinions of the astronomical world.

The Norse mythology the Vikings embraced appears to have its roots in Iceland, stemming back to before the 13th century. The original mythological scripts the Vikings would have interpreted were written in a dialect of Old Norse, which is a

North Germanic language spoken by the Scandinavian people of the Middle Ages. It's these ancient texts and poems that would have given birth to the Norse mythology we know and admire today. The mythology that has captured the world's imagination and made Scandinavia one of the most interesting and idolised places on Earth.

When the Vikings were not disputing territory nor controlling the local shipping routes, they spent a great deal of time indulging in both mythology and critical thought. From writings to drawings, many records of their ancient beliefs and theories exist. An example of one of these ancient critical thought texts is that of the *Kongespeilet* or *King's Mirror*. This text was written circa 1,200AD and doesn't propose any theories on the lights, but it does show how intelligent the Vikings of the time were. It also displays a rebuttal of the mythology that was rife during this period:

> *It is true of the northern lights, as of many other things of which we have no sure knowledge, that thoughtful men will form opinions and conjectures about it and will make such guesses as seem reasonable. But these northern lights have this peculiar nature, that the darker the night is, the brighter they seem, and they always appear at night but never by day, and rarely by moonlight. They resemble a vast flame of fire viewed from a great distance. It also looks as if sharp points were shot from this flame up into the sky, they are of uneven height and in constant motion, now one, now another darting highest; and the light appears to blaze like a living flame...*

Although the above text shows there were intelligent and analytical individuals amongst the Viking population, who understood the limitations of their knowledge and recognised there were things they did not yet know. Many others, as is evident in all populations across the planet, attested that the appearance of the aurora was unquestionably connected to the religious and mythological beliefs of the time, and not to reasons more terrestrial. Many believed that the supreme gods such as Odin, Thor, and Loki lived within a celestial location known as Åsgard, which is one of the nine worlds of Norse mythology. Åsgard was a mystical city that sat high within the clouds, surrounded by rugged landscapes and towering mountains. It was also home to the dominating fortress of Aesir. In ancient scribes and illustrations, the fortress of Aesir is often depicted as a giant powerful triangular structure made from what looks like towering church organ pipes. Everything else within Åsgard appeared miniscule under the dominating stature of the Aesir – sort of like living in an afterlife dictatorship. It makes Buckingham Palace look like a one-bedroom apartment. If a Viking lived a full and proper life, they would aspire to enter Åsgard upon their death. But Åsgard was only accessible via the precarious crossing of a large colourful bridge known as the Bifröst, which was built by the gods of old as a way of connecting heaven with Earth. The Bifröst is a majestic arch that stretches across the sky, which when depicted within drawings and mythological texts often appears to be made of a colourful glistening crystal. Due to the Bifröst acting as a freeway between heaven and Earth, the gods placed an all-powerful gatekeeper known as Heimdal at its entrance. Heimdal, who is depicted as either a commanding, fearless warrior or jolly drunken Viking who is rather rotund, possibly bibulous, and enjoys nothing more than hosting a family barbeque in a novelty apron, would judge each individual on their merits to ascertain if they were worthy of access to the paradisical lands of Åsgard. If Heimdal deemed an individual worthy, the gates would swing open, and the barbeques would begin. If not, then you would be left outside to peer through the fence and just smell the sausages.

On occasion when winter arrived in the north, the skies were ink-black, the Moon dull, and clouds absent, it would be possible to witness the Bifröst in the Arctic skies. Each time it was seen it could vary in colour, shape, and movement, but would still hold the same power and heavenly appeal. Today, we know the Bifröst as the aurora borealis. But it would be astonishing to think that each time the aurora appear in the sky, the Nordic gods and slayed Vikings of old are walking along their length, gazing down upon us from the heavens. And whilst they do, Heimdal is laughing aloud to himself whilst drinking copious amounts of mead from a rather large, engraved tankard. What a guy.

The source and representation of the aurora was not an agreed upon monopoly within Norse mythology. There were a few theories formulated and proposed, some of which overlap and others that stand alone. Very similar to the different branches of the same religions we see today.

A contrasting Nordic belief placed the origin of the auroras within the hands of the Valkyrie, who were a host of female figures that rode grand, galloping steeds in the sky. The Valkyrie, meaning 'chooser of the slain' was responsible for choosing who may live and who may die in battle. The souls of those chosen to die would be harvested by these fearsome female warriors, half of whom would be taken to the afterlife hall of slain, Valhalla, ruled over by the powerful, bearded god, Odin. The remaining half would be taken by the attractive goddess of love, beauty, gold, and death, Freyja, to the afterlife field of Fólkvangr – if I were to die a Viking warrior, I know which of these gods I would rather spend my afterlife with – subject to my wife's approval. This meant that once a warrior stepped into battle, their personal outcome was entirely out of their hands. They fought sword in hand at the whim of the Valkyrie. The story goes that as the Valkyrie rode the heavens during times of turmoil, their helmets, armour, and spears would shed a strange flickering light upon the skies. This flicker could be seen on occasion, especially when the skies were dark, clear, and exposed to the naked eye. As you have undoubtedly concluded, the light that flickered upon the skies from the Valkyrie's armour was the aurora borealis.

The actions of the Valkyrie were respected to such an extent that they too were memorialised in a text known as *The Prophecy of the Seeress* around the year 1,000AD:

> *A Valkyrie rests*
> *On the rock in steep*
> *Flickering fire*
> *Flames about her*
> *With the seep-thorn Ygg*
> *Her erst did prick:*
> *Other heroes she felled*
> *Than he had willed*

Other Scandinavians fashioned myths that were equally as magical as the ones we have already seen. In ancient Finland the aurora borealis were referred to as, 'Revontulet', which when translated equates to 'Fox Fires'. This ancient myth, a beast fable, says the auroras were created by a magical fox. This fox would sweep its tail across the snow causing it to spray up into the sky, which in turn, would create the aurora. In Sweden people named the auroras 'Sillblixt', translating as 'herring flash'. The Swedish people claimed the North Sea was so full of herring that when the moonlight shone, it was reflected by the glistening scales of the fish, causing the aurora to be cast upon the sky. The Danish believed that at night, swans would compete to see who could fly farthest north. On occasion, some of the bravest swans would fly so far north they would become trapped in the ice. They would then flap their wings in a panicked attempt to escape, which caused reflections on the ice creating the aurora. Other Scandinavians thought the lights were the souls of departed beautiful women (I don't know what happened to the ugly ones) or were caused by females who had never married cooking fish over a campfire whilst dancing. The Lapps, or Sami of Lapland, believed the lights were caused by the emitted energy of departed souls. As one can imagine, a belief such as this led to the scattering of children when the aurora appeared – I suppose to Lappish children the aurora was the equivalent of the Bogeyman. This belief formed a tradition amongst tribes that when the aurora appeared, people would behave solemnly and sit quietly in a moment of reflection. But even though the arrival of the aurora signaled a quiet and somber time for many, whistling was permitted, as it was believed that whistling at the lights would tease them closer to Earth, where they could pick you up and carry you away. This is only my opinion, but I don't see the attraction in whistling at a group of dead souls until they pick you up and carry you away. But maybe it's just me. The only individuals permitted to make any real noise during an aurora light show were the Lappish shamans. These shamans would bang a drum decorated in elaborate symbols that represented the aurora. They believed that by playing the drum during an aurora performance they would be able to harness the energy of the lights. This energy could then be used to rectify disputes and mitigate future conflict. This harnessing of power was a delicate process, however, as it was believed that if you were to disrespect the lights, they would bring sickness and possibly even death upon yourself and tribe. This rather unsettling belief of inflicted pain and suffering was passed down from generation to generation via a rather insidious story:

Biete, a young Sami boy, was wandering through the rugged Lappish Mountains with his brother, Garrell. The cloak of night had fallen, and the Arctic darkness was all consuming. But even though the Moon shone lightly in the night sky, the grip of blackness remained almost impenetrable. The reason the brothers were out at such a late hour was because they were in search of their reindeer herd. It was their duty to collect the herd and bring it back to the safety of lower ground before the true coldness of night took hold. As the brothers scoured the mountains, Biete, who was a rather disruptive Sami boy, sang a few insulting, made-up songs aloud, in order to mock the Sun, Moon, and stars. Garrell panicked and pleaded with his brother to stop. But unconcerned with his brother's plea, Biete continued to sing, and with each word, mock the heavenly creations again and again. Garrell warned Biete that if he didn't stop then evil would befall him. In an instance, a bolt of lightning shot down from the skies and killed the reindeer pulling Biete's sled. In light of this, Garrell continued to plead with Biete to stop his insulting songs, but Biete remained unfazed. Garrell pointed to the dead reindeer as proof of the anger of the heavens, yet still, Biete continued. Later that night the skies began to brighten under the presence of the Northern Lights whilst Biete continued to sing. The lights grew vivid with rage; consumed with hatred from the insulting words Biete had placed upon them. Then, in one fatal swoop, the lights struck Biete a lethal blow, leaving his body limp, dead. Garrell could do nothing but watch in silence as his brother was killed. After all, Garrell was a mere Sami boy; he was no match for the heavenly aurora. He had no choice but to take the herd of reindeer from the mountain and leave his brother's dead body behind to freeze.

The Sami tribe had many strange beliefs that transcended the generations. Some of which may help in our understanding of what it must have been like to live in such harsh terrain in ancient times, with pretty much no technology and a starkness of life that is almost inconceivable today. These stories aid in helping us understand what life must have been like hundreds and thousands of years ago living under the skies of the Northern Lights.

Many of the myths and legends of the Sami tribe orientate around various monsters and demons of the underworld. Others frequently involved a race of troll-like giants called Stallos. The Stallos trolls are modernly depicted as miserable looking, grumpy, furry, slob-like giants, with glazed eyes and terrible posture who are prone to fits of rage – like most people I know. The Stallos were notoriously stupid, constantly stumbling over trees and bumping into things. Yet somehow, they managed to just about feed themselves and fornicate. Many of the stories involved the Sami outwitting the Stallos, often with ease. But if a Stallos were to catch a Sami the results were nothing short of eye-watering, as the stupid, giant, furry troll would suck out the Sami's strength, and often innards, through an iron pipe. These stories of the Stallos, of which there are numerous, echo a very similar moral to the more commonly known tale of David vs Goliath. They allow us to delve deep into the human psyche of thousands of years ago, where even then, the people of the west and the ancient tribes of the Arctic knew strength was no match for wit. In order to exemplify this point, one such tale of the Sami tribe tells the story of how a small and intelligent man kills and beheads a Stallos, with nothing more than basic tools and a keen intellect.

Another tale tells the story of a monstrous giant named Biegolmai, or the Wind Man. The story says that in the beginning of time Biegolmai created the Sampi region with the aid of two rather stupendous shovels. One shovel he used to create wind, and the other to drop giant dollops of snow so large that nobody could live within the region. But one day the Biegolmai's shovels broke, meaning he could no longer create fearsome wind and torrential snow, which soothed the terrain and allowed the Sami people to enter Sampi and make it their home.

Furthermore, one of the most famous stories of the Sami people is the tale of the *Pathfinder*. In this story a Sami village is attacked by a vicious, marauding tribe. The tribe's people don all the weapons they can muster, but their attempts at defense eventually prove futile and the entire village, bar one small boy, is slaughtered. Once the invading tribe has taken hold of the village, they force the young survivor against his will to lead their insatiable attack to the next rural community. The boy reluctantly agrees to the delight of the invading tribe, but little do they know that his intentions are rather insidious. The small boy leads the tribe into the mountains, to a point from where the bloodthirsty savages can mount their next attack. But as darkness falls, the tribe becomes concerned, and insists they camp for the night before re-assembling at daybreak. The young boy reassures the savages he knows the mountains well, and that an attack by night will be much swifter and successful than an invasion by day. He suggests the savages tie themselves together in a line, with a large piece of threadbare rope. This way none of them will become lost under the cloak of darkness. The savages agree and are soon on their way, pleased with the fact that the young Sami boy has become a rather loyal and obedient slave. The boy then leads the tribe down the mountainside to the edge of a cliff, where he throws his torch off the edge into the darkness and shouts 'follow me!'. The savages, each tied together, follow the falling torch into the darkness, unbeknown, leaping to their death.

Many of these types of stories are still shared by the Samis today. Although most people know they are most likely fictitious and simply used to instill morals and educate the young Samis in the history and culture of their bloodline. But as technology advances and slowly encroaches upon the polar region, these stories, and the traditional way of life of the Samis, are disappearing. And the skills, crafts, and culture of the old days are regrettably being forgotten. On a rather despairing note, this advancement of technology and the level of pollution and waste it brings is contributing to the world's global warming situation. This global warming is not only melting the ice caps, causing ocean levels to rise, and forcing the tribes of the Arctic further south. It is changing the weather, and with these changes comes a reduction in aurora visibility. Therefore, the spectacular Northern Lights shows that brought much comfort, numerous myths, and stories such as the ones we have just seen, and many cultural traditions to tribes such as the Samis, are becoming less and less frequent, leading to a concerning simultaneous loss of both tradition and habitat.

Many people associated the aurora borealis with dancing. The Norwegians, Scottish, North Americans and Eskimo tribes of Greenland all made this connection, independently. The Norwegians claimed the lights were caused by the dancing and waving of the spirits of old maids. The Scottish named the lights the 'merry dancers' due to their fluid and rhythmic motion. The North American Eskimo tribes of the southern Yukon River believed the lights were caused by dancing animals, especially deer, seals, and salmon – I had no idea salmon could dance. And the Eskimos of eastern Greenland attributed the lights to the dancing of children who had died at birth. They claimed the dead children's dancing would cause the lights to be cast upon the skies.

In contrast to the above beliefs, the Young Labrador Eskimos believed the auroras were caused by torches lit by the dead. As the dead carried these torches, they were playing football in the heavens with a walrus skull; and the Eskimos would dance to the torches' rhythm. The Eskimos of Nunivak believed the exact opposite. They believed the lights were caused by the spirits of walruses playing football with a human skull. Some Eskimo tribes even believed that the most fortunate of their tribe members would spend their next life in the skies of Earth, in the 'land of eternal day', where they would partake in gigantic aurora football games. After spending some time with the local Inuit populations of these areas, this 'land of eternal day' was described in 1920 by the Danish explorer Knud Rasmussen as '…the land of glad and happy souls… [where] the people only live for pleasure.' He said it was the Eskimos' equivalent of Heaven, their hedonistic paradise. He also claimed that if you listened closely, you could hear the 'whistling, rustling, cracking sound [that] is made by the souls as they run across the frost-hardened snow of the heavens.' It can be imagined that it was this sound the Eskimos revered most, as they waited in anticipation for it to be their turn to play football in the sky. In other areas, the Makah Indians of Washington thought the lights were caused by giant fires in the far north that had been lit by a tribe of dwarfs (half the size of a canoe paddle) who were so strong they could catch whales with their bare hands. Now that's a feat I'd like to see. The Chinese believed the lights were caused by the fire of good and evil dragons as they fought a celestial battle across the firmament of Earth. The Japanese, on the other hand, believed that if you conceived your child under the Northern Lights they would be blessed with intelligence, fortune, and dashing looks – obviously my parents missed the opportunity. This list could go on and on for there are hundreds of myths as to what the auroras were and how they were formed. Doubtless, an entire book could be dedicated to aurora myths, and it would almost certainly prove ridiculously entertaining. But one thing is for sure. Eskimos and Indians didn't lack imagination, or possibly access to considerable stockpiles of hallucinogenics.

From what we have seen so far, the auroras appear to have either had a positive or a mythological, spiritual, or religious effect on the people who witnessed them. It is also noted that the tribes and people who associated the aurora with something positive appear to have lived higher within the northern hemisphere and would therefore have been exposed to the lights on a more frequent basis.

When the tribes and people of mid to southern Europe witnessed the lights, it would have been on a drastically more infrequent basis. It is said that the people of mid to southern Europe would have only seen the lights once or twice every handful of years, if at all. And when they appeared at these lower altitudes, they would tend to have been rich red in colour, which would further rally the imaginations of the spectators below. One such chap named Alfred Angot wrote in the 1500s that whilst looking at the sky he saw 'Bloody lances, heads separated from the trunk, [and] armies in conflict were clearly distinguished.' But he wasn't finished there. He continued to say that 'at the sight of them people fainted…others went mad. Pilgrimages were organised to avert the wrath of heaven manifested by these terrible signs.' He was obviously a rather loquacious gentleman. In another such incident in 1583, the auroras shone so brightly in Europe that people marched to Paris to confess their sins to King Henri III, for fear of the arrival of the rapture. However, as I'm sure we can agree, the sighting of such foreign objects in the night's sky whilst Earth was so little understood, which both appeared and disappeared without trace nor explanation, would undoubtedly have been a nerve-racking experience for all involved. It's this scarcity, amalgamated with the disturbingly potent-red colour, which attached a more negative meaning to the auroras and caused them to spread fear as opposed to comfort, reassurance, and familiarity.

In ancient Greece, the lights would barely have been seen at all, as the auroras are notably most active around the poles of Earth. Even somebody who lived a long and prosperous life may never have witnessed the lights on a single occasion. They would merely remain a myth or legend that could not be disproved. When Aristotle watched the lights in 344BC it must have been an extraordinarily rare occurrence, as the aurora borealis rarely venture so far south. Due to the light's unfamiliarity, the Greeks grew to believe that when the auroras did arrive, they brought with them the signs of war and sickness. As romantic as it sounds, to witness the auroras on a clear Greek night was not considered a moment of dancing and celebration, an opportunity to watch walruses playing football or salmon dancing the Macarena, but an instance of fear and apprehension of the terrible things to come. Similar beliefs were held by certain tribes within North America. The Fox Indians of Wisconsin believed the lights were the ghosts of slain enemies who were watching from above, waiting to take revenge on the tribe below. Within certain tribes the fear they instilled was so boundless that the Point Barrow Eskimos used to carry knives on their person to keep the lights away or stab them if they came too close (you can draw your own conclusions as to the effectiveness of stabbing light), for fear that their arrival would bring distress, pain, and suffering upon them. One famous Point Barrow Eskimo named Velma Wallis, who is still alive at the time of writing, recently said she never carried a knife to protect herself from the lights like many of her fellow tribe and forefathers. Yet she was always heedful to respect the auroras, as she was informed by her ancestors that the lights were an omnipotent ethereal spirit. She further proclaimed during her interview that an individual should never whistle at the aurora, no matter what: 'There were stories of people who had whistled at the Northern Lights...' she says, '...and they were taken [away]'. There were also nights when the lights would appear as a potent red in the sky, like fresh blood pumping through transparent arteries, and then a few days later, somebody would die, and people would say, 'The Northern Lights did that'. It seems the fear and negative stigmatisation of the lights was, and possibly still is, rife amongst many tribes of the lower northern hemisphere.

Some people associated the arrival of the auroras with individual disasters, such as war, famine, or death – which is a prevailing belief that certain folks still hold today. For example, the bombing of Pearl Harbor, Hawaii, USA by the Japanese on 07th December 1941, which destroyed 188 US aircraft, wounded 1,178 people, and killed 2,403 individuals, is believed by certain tribes to have been foretold by the appearance of the aurora in the north a few nights earlier. Over the centuries it has been believed by some that the aurora foretold the death of Julius Caesar (44BC), the death of Thomas Becket (1177) the American Civil War (1860) and the Winter War (1939).

Given the variety of myths and beliefs that have been attached to the auroras over the centuries, the fluffiest and most comforting of all belongs to the Algonquin Indians, whose name alone makes them sound like a small porcelain ornament with a fishing rod that you place in your garden. The Algonquin Indians believed that Nanahbozho, the creator of Earth, had travelled to the north upon completing our planet. He then remained in the north to watch over Earth with a studious eye to ensure all his animals and people were happy. Then, I suppose whenever he felt like it, Nanahbozho would light stupendous fires that would reflect upon Earth's surface and create the aurora borealis. It also turns out that the reason for his compulsive pyromania was rather symbolic: the fires were lit so the people and animals of Earth knew he was still there, and that he still loved them.

These myths, plus more, continued for thousands of years. As mentioned earlier, they were remembered and passed down through the generations via songs, stories, and written documents, which would have only been read and understood by the privileged literate few of the time. But regardless of how much excavation we have undertaken over the years, or the amount of research that has been poured into astronomical history and its unveiling, no exact starting point has ever been noted from which these myths sprouted. But it is possible to trace the beginning of their end.

Let's jump forward to 1592.

THE BEGINNING OF THE END

'Sometimes what you think is an end is only a beginning.'

Agatha Christie

In 1592 as the Japanese invaded Korea, on the other side of the world in, Champtercier, France, a small boy named Peirre Gassendi (1592-1655) was born. In each of his depictions Gassendi is shown as having rather bulbous eyes whilst donning a little smile that makes him look like a disruptive, mischievous child who has just done something he knows he wasn't supposed to. At an early age, Pierre showed himself to be somewhat of an academic prodigy, and through a combination of natural intelligence and academic rigor, he secured himself a position at the college of Digne, where he demonstrated an aptitude for languages and mathematics. After his studies at the college of Digne were completed, he moved onto the University of Aix-en-Provence where he studied philosophy. After completing his studies in 1612, the college of Digne contacted Pierre with a request for him to return to the establishment and undertake a lecturing position within Theology. Pierre accepted, and shortly after this invitation he officially became a man of the church. It didn't take long before the University of Aix-en-Provence re-established contact with Pierre, however, and offered him different lecturing position in philosophy. Driven by his overriding passion for philosophy, Pierre accepted the new position, left the college of Digne behind, and undertook his post at the University. As a lecturer, Pierre's topic of choice was Aristotelian philosophy. It is somewhat ironic, then, that Pierre Gassendi would later be remembered for his philosophical works and observations of the auroras, just as Aristotle had been when he witnessed the lights in 344BC, roughly 1,900 years earlier. From this point on, Pierre would become involved in following the work of both Galileo and Kepler; an interest that would inevitably lead him towards a first-hand experience with the Northern Lights. At this point in history, science was still the underdog in society, and its theories and intrusive ways were not fully accepted by the people. The reason being was that religion was still the dominating force of the time, and the nosey, inquisitive scientists of the moment were finding holes in the scripture and picking apart its belief systems. A point in case can be provided from the findings of the Polish mathematician and astronomer, Copernicus. With his floppy-hair and chiseled-chin, Copernicus hypothesised that the Sun was the centre of the Universe, not the Earth. This belief evidently struck a chord with Galileo who became an avid proponent of the idea. But Galileo's combination of public influence and scientific advocacy ineluctably led to him being arrested, tied-up and made to declare under duress that 'I, Galileo... swear that I will never again say or assert that the Sun is the centre of the Universe and immovable, and that the Earth is not the centre and moves'. Evidently, on this occasion, Galileo was not a man of his word, for the inquisitive trudging of science plodded on, much to the detriment of religion.

In 1621 at age 29, Pierre would encounter the Northern Lights and all their beauty for the first time. It could also be said that Pierre was a serendipitous man who had been born in the right place at the right time, as the first 25 years of the 17th century have been noted as one of the most active periods of aurora activity in history. This serendipity provided Pierre with an opportunity he would never have received if he had been born 50 years later or within the southern hemisphere. As once we reached the year 1625 the Northern Lights rarely shone again, as they slipped into a lull that would last almost 100 years. Consequently, chance plays a role in our understanding of the auroras.

During this year, Pierre would watch the lights dance in the night sky just as the Eskimos, Indians, Vikings, and his strongly influencing predecessor, Aristotle, had done. It also turned out that just as the lights had left a lasting impression on each of their spectators over the thousands of years gone by, Pierre was no different. Just as it was with Aristotle, Pierre's experience with the Northern Lights was so influencing, he decided to document it.

Pierre had already proven himself within the field of science and had been noted as one of the first people in history to adopt a modern 'scientific outlook'. Throughout his life he celebrated many achievements. He had gauged the apparent diameter of the Moon; written on the rotation of Earth; measured the speed of sound to an accuracy of 25%; and suggested through experimentation that the creation of a vacuum was possible – something that was later proven correct during a laboratory experiment in 1643, 12 years before Pierre's death. This meant that when Pierre wrote on the Northern Lights his words were both valued and accepted by the scientific community.

After witnessing the lights in action, Pierre corresponded with multiple observers who were located as far away as the eastern Mediterranean. His own observations, coalesced with the recordings of others, led Pierre to hypothesise that the source of the lights was contained at a very high altitude above the North Polar Region. His hypothesis was not entirely correct, but he was significantly closer than anybody else had ever been throughout the entire of documented human history. He believed the cause of the lights was the same thing that caused Parhelia, or Sun Dogs, where the Sun's rays penetrate ice and snow crystals at high altitudes and form a luminous halo around the Sun. He also realised that no official name had been given to the hypnotic, dancing lights he witnessed, and therefore took on the responsibility of providing them with an appellation. He named them, *aurora*. Pierre chose this name as Aurora is the Goddess of Dawn in Roman mythology. As the sister of Helios and Selene, the goddesses of the Sun and Moon, Aurora would rise from the sea each morning riding a horse-drawn chariot across the sky. She would fly up high before swooping down gracefully in an arc that allowed her to trace a line across the face of the Sun. Within her hand she would carry a large pitcher, which she would use to shower water over the surface of Earth and onto its inhabitants below. Then, as the Sun grew tired, she would begin the transition of blending day into night and then night back into day. Therefore, *aurora* must have seemed like an entirely apt name for a series of lights that dance in the sky and bring light to the darkness of night.

Pierre Gassendi's theory on the lights reigned supreme as the most accurate explanation of the auroras at the time; and the work of other physicians such as Benjamin Franklin and William Gilbert further complimented and reinforced his postulations. Gilbert proposed a hypothesis that Earth was a giant magnet and that this magnetic property may have influenced Earth's interaction with its surroundings. Another physicist, Sir William Cooke, later took this idea even further. He constructed an experiment whereby he isolated gas within a vacuum tube, and then transmitted electrons through the tube. What he discovered upon doing this was that the gas contained within the vacuum would glow. He also learnt that depending upon the type of incandescent gas he placed into the tube, when he pumped the electrons through, the varying gases would glow different colours. He then placed a magnet at the side of the vacuum and moved it around to see what would happen. What he found was astonishing. He learnt that the electrons were either attracted or repelled by the powerful magnet as he held it close to the vacuum, meaning the illuminated gas beams would either move towards or away from the force. It was a profound piece of scientific research. But given how spectacular this discovery was, he had not actually solved the riddle of the auroras (this would have to wait for the coming of the next century) but he was playing with the right tools and was forming the right ideas.

All-in-all, at this moment in history, unbeknown to the scientists of the time, the underpinning of the understanding of the auroras was slowly beginning to emerge. Research and evidence were steadily starting to accumulate in four distinct areas that directly related to the lights: The Sun, Earth's magnetism, atmospheric gas, and electricity. However, the link between these discoveries had not yet been tangibly tied together, but the foundations of understanding were gradually being laid. As it turned out, Gassendi's theory stood strongly for 120 years until two Swedish astronomers, Olof Hiorter and Anders Celsius (famous for creating the Celsius temperature scale) came along and proposed something closer to the truth, but still, not quite correct.

Celsius and Hiorter were both born around the turn of the 18th century, with Hiorter being the older by a couple of years. Celsius studied astronomy at Uppsala University, Sweden, whereas Hiorter undertook his studies in astronomy in the closely neighboring country, Netherlands. The two would never have crossed paths if it were not for Celsius' decision to vacate his position as lecturer at the University of Uppsala and undertake a grand tour of European observatories – predominantly those in Germany, Italy, and France. The position that Celsius left behind was soon to be filled by Hiorter, who would lecture at the University and then later meet Celsius upon his anticipated return. Over his years as a studying, lecturing, and practicing astronomer, Celsius made the decision that he wished to study the auroras that were so often witnessed towards the Northern Polar Region. Whether he had heard they impregnated people, attacked innocent Eskimos, and played football with people's heads is unclear, but he was attracted to the prospect of understanding them further, nevertheless. He then chose to offer the position of assistant to his Uppsala University replacement, Hiorter, who accepted the opportunity to try and dodge impregnation with haste.

Between the years 1716-1732, Celsius and Hiorter recorded 316 individual observations of the aurora borealis. Their observations must have led to many new and astounding conclusions because up until this point, nobody in history had ever taken such a rigorous interest in the auroras, nor assessed them from such a scientific standpoint. The overriding conclusion that Celsius and Hiorter drew was that there was a strong connection between the aurora borealis and the magnetic field of Earth. This was entirely correct and proved to be a humongous step forward for astronomy across the globe. Celsius arrived at this conclusion by noticing that when the auroras appeared there would be unusual movement within his compass needle. And the stronger the aurora activity was, the more violent the movements of the compass needle would become. This exact idea was also being pushed at the time by the astronomer, Edmond Halley. But it appears that Celsius and Hiorter were the ones with the greater substantiated evidence and were therefore the ones most adequately equipped to push the idea upon society. Curiously, this bizarre connection between a compass needle and the auroras had been noticed and reported in the past by numerous Scandinavian and Russian sailors. But, as the sailors of the time tended to be drunk, uneducated, untrustworthy, and deemed capable of nothing more than swabbing decks and singing shanties, these reports were dismissed on the spot. It wasn't until somebody considered more credible brought forward the same evidence, such as Celsius, that it was taken seriously.

Just as Celsius and Hiorter's discoveries were being published and celebrated by science in 1733, a French astronomer, geophysicist, and chronobiologist who went by the name Jean-Jacques d'Ortous de Mairan, was pushing for his own theories on the lights to be accepted – which he had formed 16 years prior in 1717. Jean-Jacques, who is rather plump with a crooked nose and long face, was the first person in history to attach the materialisation of sunspots to the forming of the Northern Lights. His theory was entirely correct, and although he was certain of it, he lacked significant proof, meaning neither he nor anybody else knew for sure he was right. What's more, it appears Jean-Jacques was rather disliked, and his work didn't present itself as pleasant reading. This was possibly because he named his research paper the *Dissertation on the cause of light phosphates and noctilucence or nightly light* – a title that doesn't exactly tease the reader into wanting to know more. But the fact nobody liked nor agreed with Jean-Jacques' theories didn't seem to matter. What did matter was that both Jean-Jacques' and Celsius and Hiorter's discoveries and publications instigated debate and discussion. They thrust the auroras onto the public stage. And it is thanks to this that we know what we do today. Finally, the scientific quest to understand the auroras had begun.

This scientific pursuit was the perfect accompaniment to an already existing thirst for exploration that had swept society and preoccupied the minds of the public, elite, and intellectuals alike. Further studying and discovering the world was somewhat of a pastime for many of the wealthy, who would often employ individuals to undertake flamboyant exploratory quests in their name – a hobby that is overdue a resurgence in my opinion. It became such a prestigious sideline of the well-heeled that many took to it with great vigor. It was also a major task that had been accredited to the military of the day, especially the Royal Navy. And there is one rather famous chap from this military establishment that dips his toe into our story, who I am almost certain you have heard of at some point. The name of this gentleman: Captain James Cook.

Whilst in the midst of a rather distinguished military career, Cook graced many foreign shores and travelled much of the world under the authority and expense of the government and monarchy. He had mapped the entrance of Saint Lawrence River, charted unknown seas and reefs, spent years sailing within the Pacific and mapped the coastal lines of New Zealand, Hawaii, and many countries in Europe to a detail that had never been known before. He was a true craftsman, a stoic explorer. But in the year 1768 Captain Cook was tasked to head towards the 'Great Southern Land', known as present day Australia, to undertake a voyage of epic proportions. His commission was to investigate and chart these comparatively unknown territories and claim their shores for the monarchy. He was further tasked with undertaking precise scientific measurements of Venus's trajectory across the skies of Earth for the benefit of the Royal Society – something he messed up spectacularly, which must have been unimaginably frustrating after sailing for months on end to gain the world's best vantage point, only to expose that his maths wasn't quite as good as he told his boss it was.

After such a successful expedition to the south, once returned home to the UK, Captain Cook was deployed to the Pacific again, with another pioneering assignment to complete. It was on this voyage on 17th February 1733 that he would witness, and document, the existence of the auroras within the southern hemisphere. This was the first time Cook had ventured south of the Antarctic Circle, and unbeknown to him, he was heading for a visual display akin to nothing he had ever witnessed before. Saying this, as Cook was a man of science, he was fully aware of the lights' existence in the north and had even read the works of Peirre Gassendi on the Northern Lights that had been published a few decades earlier. Evidently, he also gave a moment of thought to the writings and works of Peirre, as after he had witnessed the lights' display in the Antarctic Circle, he adopted Peirre's theme for naming and christened them the aurora australis, or Southern Lights. This sighting, according to many academics, is considered to have been the first reliable observation of the Southern Lights outside of the already existing aboriginal tribes, which up to this point had been the sole occupants of Australia for thousands of years. But although Captain Cook is accredited the glory of the first reliable sighting of the aurora australis, it seems, as is often the case, the real kudos should belong to a rather lesser-known individual, William Wales. On the night of 17th February when the Southern Lights were first witnessed, William was on deck, fulfilling his role as the ship's designated astronomer, and it is his notes that marked the passing of the lights and still survive to this day. William wrote on the lights that:

> *I... found it to be the very same phenomenon which we call the Northern Lights in England. The natural state of the heavens, except in the S.E. quarter, and for about 10° of altitude all round the horizon, was a whitish haze, through which stars of the third magnitude were just discernible. All round, the horizon was covered with thick clouds, out of which arose many streams of a pale reddish light, that ascended towards the zenith. These streams had not that motion which they are sometimes seen to have in England but were perfectly steady, except a small tremulous motion which some of them had near their edges.*

But it seemed that William's sighting of the auroras was not to be an isolated incident – something that must have been an exhilarating experience for an astronomer watching the skies on the geographical fringes of civilisation. But if it was, he certainly didn't show it in his diary, because in the extract below, we see that mid-performance he decided to go to bed:

> *About 10 O'Clock a Phenomenon appeared in the heavens in many things resembling an Aurora Borealis but different materially in others: it consisted of a dull reddish light reaching in hight [sic] about 20 degrees above the horizon: its extent was very different at different times but never less than 8 or 10 points of the compass. Through and out of this passes rays of a brighter colour [sic] light tending directly upwards; these appeared and vanished nearly in the same time as those of the Aurora Borealis, but were entirely without that trembling or vibratory motion observed in that*

Phenomenon. The body of it bore from the ship SSE: it lasted as bright as ever till nearly 12 when I went down to sleep but how much longer I cannot tell.

It was later written in the ship's notes that the auroras had not just been witnessed once or twice, they had danced among the stars on at least nine consecutive nights as the crew drifted peacefully within the Southern Ocean. The observations of the lights were also catalogued alongside further accounts that told the story of the uninhibited skies of the Antarctic, where it was written that the night skies were extraordinarily lucid and were often littered with thousands of stars and visited by fleeting meteors of 'uncommon brightness'.

After these observations in 1773 became public knowledge, the Chilean writer J. I. Molina affirmed the lights' name and sightings of that night and wrote, in 1810, that the lights had indeed been seen 'towards the Antarctic by the well-known [Captain] Cook and other educated sailors' and that 'unlike the others [the lights of the northern hemisphere] these [sightings] were [named] the Aurora Australis' just as Captain Cook had intended. Obviously proud of what he had seen and achieved throughout his past voyages, Cook conceitedly stated that 'by twice visiting the Pacific Tropical Sea, I had not only settled the situation of some old discoveries but made there many new ones and left, I conceive, very little more to be done'. How wrong he was.

Since these discoveries three centuries ago, we have taken their work and researched the auroras in greater and greater detail. A compilation of data has been formed around them, all generated by scientific researchers such as astronomers, physicists, and astrophysicists. Indeed, for many of the years that followed Celsius and Hiorter's discoveries, the world's understanding of the lights lay somewhere between myth, science, and hypotheses.

Over the years that ensued, many theories were tried and tested, and many myths dispelled. But, eventually, after much rigorous study, more elaborate conclusions were drawn, and greater piles of evidence were formed. One such fellow who contributed more significantly than any other individual to our understanding of the auroras was the troubled Kristian Olaf Bernhard Birkeland. His story is intertwined with the lights, so much so, that any conversation about the auroras would not be complete without him. Entire books have been written on his achievements, ideas, and contributions to science and understanding. But for the purpose of our story, we will contain his plethora of achievements into a single chapter. We'll also keep it casual and call him Kristian.

THE CURIOUS LIFE OF KRISTIAN OLAF BERNHARD BIRKELAND

'There is no great genius without some touch of madness'.

Aristotle

Kristian was born in 1867 – 134 years after Celsius and Hiorter's publication, and 150 years after the public started ignoring Jean-Jacques. As a prodigal Norwegian with a superbly floppy moustache and circular thin-rimmed glasses, he was a curious character of small stature (only 5ft 5 inches tall) that exuded a general air of quirkiness and an obsession with all things magnetism. He was an avid inventor and ideas man who wrote his first scientific paper at age 18, which as we can guess, was orientated around the study of altitude, atmosphere, and the Earth's magnetic field. During his short yet fruitful life, Kristian achieved much, slept little, battled with depression and numerous ailments, overdosed on drugs, lived in a state of paranoia, became a reputed world-class scientist, was nominated for the Nobel Prize seven times, annoyed his wife, electrocuted himself on numerous occasions, loved whiskey, lived in the desert, blew things up, wrote books, became the first person to propose the idea of splitting an atom, took his own life, and as we shall come to see, unlocked the secret of the auroras. But even though his passion for the Universe and the way it interacts with Earth's magnetic field began very early in life when he was just a child, for the purpose of our story we will join Kristian on 14th October 1899 when he, along with three assistants and fellow scientists, left the small Norwegian mining town of Kaafjord to embark upon a journey to the top of Haldde Mountain, located within the Norwegian Lappish territory.

What Kristian planned to do was spend an entire winter on the summit of Haldde Mountain, almost 1,000 meters above sea-level (something which many, including the local Lappish community, considered preposterous) and study the Northern Lights in greater detail than anyone had ever done before. That summer he had commissioned the building of a stone house on the summit of the mountain, which would later become the team's base and home for the ensuing winter. So that day, as the wind blew at the shuddering temperature of -25c, the team slowly made their way to the summit, with the assistance of a few reindeers that pulled the sleds of gear required to live and monitor the lights during the darkest, coldest months of the year. It would have been a rather strange sight to behold seeing four men clambering their way north to a desolate Arctic peak, whilst the local Lappish community headed south to avoid the coldness and unpredictability of a northern Norwegian winter. The Lappish thought the team crazy, and in all truth, none of the men knew what awaited them at the summit, or what the winter would bring. It was a brave and brazen expedition, all in the name of science and national pride.

All-in-all, the team's inventory (minus consumables and other small items) consisted of two reindeer skin hats, two reindeer skins, skis, five reindeer jackets, four pairs of snowshoes, two pairs of Sami shoes, medical chest, flag, two mirrors, alarm clock, two buckets, three lamps, five pillows, four sleeping bags, ten blankets, three mattresses, kitchen utensils, stove, five pairs of skin mittens, tools, copper nails, two sleds, three reindeer reins, cups and glasses (for wine and beer, of course) tents, six chairs, two office chairs, and two chaises lounges – as you do.

After many hours slogging, climbing, and fighting with the skittish reindeer (which none of the men had much experience dealing with) the exhausted and frozen team arrived at the elusive summit just in time for darkness to set in. As they hurriedly unpacked their equipment from the reindeer sleds into the newly erected building that would act as their home, laboratory, and observation tower for the coming months, the auroras bloomed into existence, almost as if it were a welcoming performance for the party and a taste of things to come. Once inside the building the team worked quickly to set-up their equipment, lay-out their belongings, and make this cold, isolated, desolate stone building as homely as possible. It was by no means luxurious with its draughty cracks (no jokes please) and rickety windows, but it was all they required for the job at hand. It was secure, robustly built, and could hopefully withstand an infamous Arctic winter. Kristian was excited for what was to come as he had many plans, ideas, and aspirations, and there was much he wanted to achieve. Amongst his list was his desire to shine some light on where the auroras came from, how they were made, and what they were. He also wanted to test what I would call some of the more 'entertaining' fables, such as: *were the auroras created by mountains? Did they crackle like bubble wrap? And did they sometimes swoop low enough to touch the ground and singe your hair if you happened to be in the way?* Kristian also knew the Lappish thought the lights were prone to angry fits if noises were made during their presence. And if the local people thought they had in some way offended the lights, they would chant a rhyme to pacify them:

> *The northern light, the northern light*
> *Flickering, flickering,*
> *Hammer in its leg,*
> *Birch bark in its hand.*

So, each time the lights shone the party would remain silent out of respect – and possibly out of the fear of receiving a singed scalp and going home bald.

Kristian and his team set-up numerous instruments to monitor the lights, many of which were the most technologically advanced of the day, which allowed the team to record things never before documented. But Kristian's pride and joy were the magnetometers, which were used to measure fluctuations and changes to the Earth's magnetic field. For a while now, as we have seen, the auroras had been associated with the magnetic field of Earth, so the potential for new discoveries that these pieces of equipment brought was supremely exciting. In order to ensure nothing tainted the machines results, the room in which they were kept had to be specially prepared with a delicate touch. First of all, each of the men had to remove all metallic items before entering the room (Kristian himself had removed all the metal buttons from his jackets and replaced them with ones made from reindeer bone) they also had to don a rather fashionable pair of reindeer slippers to minimise disturbance and maximise style. The team even removed door hinges, pipes, and nails, and replaced them with copper and brass alternatives. They were leaving nothing to chance, and they took all precautions to ensure their results were not contaminated, which would reduce the possibility of contestation upon their publishing when they returned to the south. Once the machines were fully set-up, a recording timetable was established. Each man was to contribute to the work to ensure no readings were missed, and that everything was documented at the designated time. Sometimes when the auroras were particularly bright, readings were taken every 30 seconds throughout the entire night. It was a tedious job, but I suppose when you're stuck on a mountain during the depths of Arctic winter, *what else are you going to do with your evenings?*

Day by day the team became more settled and adjusted to their new-found landscape and home. It was a barren place, but none of the men complained as they each knew they had a job to do. They were here to make one of the biggest scientific breakthroughs of their time, and they were all proud to be part of such a prestigious team. As winter encroached, the Sun breached the horizon for less and less time. Each time it rose it would pop its head over the horizon a little less, until one day, the team stood and watched a full day's sunlight that lasted only nine minutes. Then, as it dropped below the horizon and disappeared out of sight, the team accepted it was the last time they would see the Sun for roughly five months. Things had just become very serious. They were so far north that the winter was an unabated perpetual night, and they knew that for months on end they would have to meticulously operate in almost complete blackness. All they would have to penetrate the darkness would be the light of the Moon, the auroras, and that emitted from the three lamps they had brought with them. Each of them was slightly unnerved by this fact, but this is exactly what they had come for. Their true scientific research could now begin.

When the Sun doesn't rise, and all 24 hours of each day are eternally black, it becomes notoriously difficult to establish a routine and comprehend what time it is. Kristian had always suffered with sleep deprivation anyway, so I suppose not knowing what time it was made very little difference to him. He was also a workaholic, so being completely isolated with no distractions was probably welcome news. He could fully commit himself to his research with no distractions to hinder his work. But as the darkness took hold, Kristian's other senses heightened as a means of compensation. Although he was not aware of the time, he was constantly aware of the wind and noises the winter provided. Sometimes the weather became so terrific he would jot entries about it in his diary. He wrote on one occasion that:

The wind sometimes roars so against the house that you would have thought that you were sitting at the foot of a waterfall; and the floors tremble and everything shakes. We are able to gauge the storm outside by the noise within. Often we cannot get out of the house ourselves for several days and it takes three strong men to shut our little door. One strong anemometer was blown apart in the course of a few days and we found pieces of it 50 to 100 meters from the place it had been put up.

He continues to say:

We have seen a layer of snow a meter thick and so hard you could jump on it without sinking in, practically disappear from the summit in the course of a few hours.

> *It may be imagined what a whirling and drifting there is in the wind, and when the snow is comparatively fresh and not pressed into a compact mass. Even indoors the situation is not always comfortable. Water freezes a couple of feet from the stove and the lamp is often blown out on the table in the middle of the room, although in a general sense the house is well-enough built.*

In my mind it must have been quite a torturous place to spend months on end, but these were hardy men with an objective, so they gritted their teeth and dealt with the situation at hand.

As the auroras shone, the magnetometers the team had set-up would record their activity, even if the skies were cloaked in clouds and the auroras could not be seen. These devices measured three things about the magnetic field and what it was doing. They measured the direction the field was moving in, the field's horizontal strength, and the field's vertical strength. This information could then be used to map both the field and the auroras, to help portray a three-dimensional vision of what was happening above. These devices were so advanced for their day that they could measure changes in magnetic variation up to 100,000 kilometers away, which as we can imagine, made Kristian a very happy man. Whilst undertaking the research, Kristian noted something quite profound. He noticed the auroras would only appear when there was magnetic disturbance in the air, and the greater the disturbance, the more powerful the auroras. This was one of his greatest realisations that would later contribute to the formation of his aurora theory. However, even though this was a magnificent discovery, it gave birth to a burdensome question: *if the auroras only appeared when there was magnetic disturbance, then where did this magnetic disturbance come from? What caused it and why?* There were not many theories about this at the time as it was a greatly misunderstood and little-researched topic, but there was one thing that was taken as scientific 'fact': Earth was a sole entity that could only be affected by itself. The scientists of the day believed that the on-goings of our solar system had no effect on Earth. It was as if Earth was in a protective case, which nothing could penetrate, and was subject only to its own laws, rules, and actions. The idea that something in our Universe could affect the Earth's magnetic field, or Earth itself for that matter, was nothing short of hilarious. And the few that dared propose this idea in the past were laughed at and ridiculed. But Kristian didn't mind. He didn't seem to care much about what other people thought as he was a true scientist – and often lashed on whiskey. He would continue to undertake his research regardless, as accurately as he could, and he would report his findings despite public scientific opinion. One thing that Kristian's research did establish, though, was that there was no connection between aurora activity and weather. In earlier years it had been believed that the auroras could be used as a sort of weather forecasting tool, letting the locals know what they should expect over the coming days. For instance, the further south the lights shone the milder the weather was expected to be. And the same was said if they were brighter. This meant that very bright lights that shone quite far south were a solid sign that mild weather was on its way. The colour red meant snow or rain were due. If the lights shone from the northwest to the southeast, then it was believed to soon be clear and cold. And if they moved a lot, it was believed it would be windy – almost as if the auroras were being blown by the wind high up in the atmosphere.

Up until this point the team's work had proven to be quite fruitful. They had seen the lights tens of times – sometimes for hours each night. They had monitored their movements, noted their possible connection to the weather, seen how they appeared to be linked to the actions of the magnetic field, and dispelled the beliefs that they sometimes touched the floor, made noises or were emitted by mountains. But on 11th January their good fortune was to change for the worse, as a fierce storm unlike anything any of the team had seen before gathered on the horizon. It didn't take long before the gigantic storm closed in, and as it did, they knew they were in for a rough time. Try to imagine being on a remote outpost in the Arctic, on top of a mountain in constant darkness, with no help or assistance nearby. You are effectively stranded in a small stone building in temperatures as low as -30c, and the biggest storm you have ever seen is ominously rolling towards you. It's at a time like this that you would wish you had packed more toilet paper.

As the clouds encroached upon the team even the reindeer knew something bad was on its way, so they chewed and pulled against their harnesses in a fretted attempt to escape. Once free of their shackles they promptly ran for the security of lower ground without looking back, with their feet slipping on the ice as they attempted to scramble as quickly as they could over the jutting rocks of the mountain peak. As they ran, the bells on their necks jingled hauntingly, as they bolted into the distance to save their own lives from the wrath of mother nature. As the reindeer disappeared down the treacherous mountainside the team began to hurriedly nail planks over the windows of the hut, whilst bringing all the food they could carry indoors just in case the storm proved stubborn and took a few days to pass. They dismantled all the aurora measuring equipment that sat outdoors and brought it inside to prevent it from being damaged, with the exception of one anemometer, which they left on the roof so they could measure the storm's wind speed. It wasn't long after they had disassembled their equipment that the beginnings of the storm began to rage, and as it did, the team bedded down in the hut under their reindeer blankets and hoped for the best. They had sealed the doors and windows as best possible, but the hut still proved draughty, noisy, cold, and uncomfortable. They could hear the wind screaming outside as it whipped past the building and caused the doors and panels they had secured in place to rattle furiously. They also had no means of contact with the outside world so all they could do was wrap up, ride out the storm, and hope the building held strong; because if the building fractured, collapsed, or lost its roof, the team were certain that death was the only outcome. They knew that if they tried to make their way off the mountain or attempted to see the storm through without adequate protection, they would likely be killed by exposure. So, they held tight, more than likely said their prayers, and hoped the storm would pass relatively quickly. But, little did they know at this point that the storm would rage ceaselessly for three weeks, without respite: their ordeal had only just begun.

For the three weeks the storm raged, sleep was torturous, the temperature inside the hut proved scarcely bearable, and the team had to physically shout at each other to be heard above the wind. Snow began to make its way into the hut through the crevices and holes in the building, where it blew into the corners and formed small drifts around the hut. The roof creaked as the steel guy ropes that held it in place strained against the wind. It didn't take long before the boards that were shielding the windows and doors began to tear away, leaving greater holes for the snow to pile through. Once the boards were dislodged there was no way of reattaching them, as they simply disappeared with the wind into the darkness. Some attempts were made to cover the fractured windows with reindeer skin, but the improvement to their condition was minimal, and almost not worth the effort. In a rather surprising turn, after a few days the men became somewhat accustomed to the weather. Albeit they were cold, sleep deprived, and constantly aware of the danger. But they settled into an uncomfortable routine that helped with the forced solitude they were enduring. But it didn't take long for things to change. One night around the time the team was attempting to settle into bed, one of the steel guy ropes that held the roof in position snapped. It cracked with such volume that the team stopped in their tracks and heard it lash against the side of the building like thunder. With one of the guy ropes now flapping loose outside, slapping continually against the stone wall, the team knew the security of the roof had been breached, and if other ropes followed suit, each now bearing additional strain, then sheer chaos was guaranteed to ensue. As if this wasn't enough for the men to endure, the front door to the hut soon ripped open and was almost torn free from its hinges as the full force of the storm rampaged into the hut like a bull. Kristian leapt from his bed whilst donning his trusty woolen pajamas and ran to the door to close it back. But due to the unrelenting strength of the wind, Kristian was unable to close the door by himself, and it took three men pushing with all their might to secure it. Once closed the team nailed the front door shut, leaving them truly locked inside the hut. If something happened now, such as the roof collapsing or the blasts of wind that flew down the chimney causing the fire to spread into the room, there would be no means of rapid escape and the men would surely die. I'm sure the sense of irony of potentially burning to death in the middle of a sub-zero Arctic winter was lost on none of the men, but they had no choice but to secure themselves in the hut almost permanently.

On occasions when the weather would partially relent, some of the team would force open the door again and run outside to try and make magnetic and atmospheric measurements. Regardless of the weather the team did not lose sight of their task and continued in the pursuit of their desire to understand the auroras. On one occasion, one of the team members took a weather kite outside in an attempt to make a collection of measurements during a lull in the storm. But unpredictable as the storm was, within moments the wind was back, and the kite dragged along the skyline. As the gust of wind was so strong, and appeared from nowhere, the individual was hauled through the snow and over a cliff edge to where he only just managed to grasp the jutting rocks a few meters down, which he held onto for dear life. If he had not been lucky enough to grab onto the protruding rocks, he would have fallen roughly 40 meters down a cliff face to his death. As soon as Kristian noticed what had happened, he tied a rope around his waist, tethered it to the front door of the building and ventured outside to save his stranded colleague. The wind whipped around him as he made his way with another colleague towards the cliff edge, to where he would soon find his friend holding on with all his might. In the short time he'd been stranded his lips had turned blue, his hands become frostbitten, and his eyes were starting to swell. Without a moment's hesitation the two men reached down and grabbed their stranded colleague's arms, hoisted him over the lip, and managed to carry him back to the hut where he was dumped in front of the fire until the colour of his lips returned and the frostbite on his hands began to subside. He was lucky to be alive, and for all the men, this incident proved to be a stark reminder of the very present danger this part of the world contained. All it took was to be in the wrong place at the wrong time and the odds would be immeasurably stacked against them.

Being trapped inside because of the storm was not all bad. For one thing, it gave Kristian time to analyse his results, draw some loose conclusions and put together the findings in his mind. He could feel the auroras starting to make sense, and he began to see how his theories and research were beginning to fit together. He was now convinced that the disturbances within the magnetic field were caused by electric currents, which he believed were not formed by Earth, but somehow arrived at Earth from elsewhere in the solar system or Universe. With this in mind, Kristian asked one of the team to draw a map of Finnmark (the location they were currently in) and sketch onto the map a series of arrows that would highlight both the direction and strength of the magnetic storms they had measured. They soon discovered that the magnetic storms appeared to run at right-angles to the magnetic field. And as we have seen already, Kristian had hitherto concluded that the auroras and magnetic disturbances were connected with one another. But with the addition of these later findings, he would rightfully conclude that the magnetic disturbances were caused by a series of electrically charged particles that were emitted from the

Sun, called cathode rays. Effectively, he postulated that the auroras were caused by solar activity. He further noted that these rays were emitted from sunspots, which were intense areas of activity on the Sun's surface. He continued to state that the reason there were not always auroras when there were sunspots was because sometimes, the cathode rays emitted by the Sun would miss Earth completely. This was one of the reasons why the auroras were unpredictable – as it is not possible to know which of the rays would hit Earth, and which would miss. Something else remarkably exciting he discovered was that similar auroras would tend to appear at 27-day intervals. As an initial thought this seems quite ludicrous. *Why would auroras sometimes almost repeat themselves, and why every 27 days?* Kristian's answer was simple: The Sun took 27 days to rotate. Therefore, the rays released from a sunspot would hit Earth, and then as the Sun rotated, the sunspot's emissions would miss us, until it had completed a full rotation meaning the emissions from the sunspot were pointing towards Earth once more. This was also a profound thought that was entirely new to science. And as Kristian tended to be, he was entirely correct.

Inspired by these discoveries and the findings from his Arctic expedition, Kristian returned to his laboratory in Oslo, Norway, determined to recreate what he had learnt through a controlled scientific experiment. Much to Kristian's disappointment, however, the measurements and observations collected during his 1899-1900 expedition were not sufficient to constitute a solid body of evidence. This left his theories and ideas unproven and open to academic harassment. One such body that considered Kristian's findings to be somewhat lacking was the Royal Society of Science in the United Kingdom. Upon reading his research the society simply stated in retort that:

There is absolutely conclusive evidence against the supposition that terrestrial magnetic storms are due to magnetic actions of the sun; or to any kind of dynamical action taking place within the sun, or in connection with hurricanes in its atmosphere… [and that] the supposed connection between magnetic storms and sun spots is unreal, and the seeming agreement between the periods has been mere coincidence.

Unfazed by this outright dismissal of evidence, Kristian leapt back to work, and two years later, in 1902, had organised another more elaborate expedition to the Arctic regions. This time he was determined to simultaneously measure the auroras from four separate points within the Arctic: Iceland, Norway, Svalbard, and Novaya Zemlya, Russia. And once these separated measurements had been taken, they could then be compared, allowing Kristian to form a truly global perspective of the aurora borealis. He also wrote many letters that would later be posted to numerous foreign observatories, asking them to give him any findings they had on the auroras to help further reinforce his conclusions. So, he set his mind on the job and knew if he could achieve this task, his findings would be beyond contestation and his name would go down in history. He was also arguing quite frantically with his wife at this point, so another winter in an Arctic hut probably sounded quite appealing.

It was on 1st July 1902 that Kristian boarded a train destined for Sweden, where he would then continue onwards to Russia. A few members of his research team had already departed for Novaya Zemlya a couple of weeks earlier, and now they were making their way back from the desolate location to the forest encompassed city of Archangelsk to meet him. Once they met, the men, plus Kristian, planned to take a seven-day journey by steamer to Novaya Zemlya. Kristian had enlisted the help of a few skilled scientists, whom he believed would be trustworthy individuals that could be relied upon to collect and document the actions of the aurora borealis, which could then be used to reinforce and support his initial findings from his first expedition. These scientists had further recruited two assistants to help share the workload and ease the expedition's burden. It rather comically transpired, however, that both assistants had been recruited because it was believed they could speak fluent Russian, which was most certainly not the case. Both had proclaimed they could to get the job, which once secured they planned to muddle their way through and hope nobody noticed their evident lack of linguistic capabilities. The first assistant, Johan Koren, had slipped a Russian phrasebook into his pocket that he would consult when necessary. When no one was looking he would study the phrases in the book in the hope he would quickly learn enough to speak when summoned. As ridiculous as this sounds, his efforts were vastly more impressive than those of the second assistant, Hans Thomas Schaaning, who convinced himself that not speaking Russian would not be a problem, nor would he try to learn it. He would simply cook good food for the group in the hope that if they all had full bellies, they wouldn't care – and he was right. It turns out the ability to whip-up a good sandwich can get you out of all sorts of predicaments. With his team established and together, Kristian's intention was now to make the final leg of the journey to Novaya Zemlya where he could check up on the Russian observatory to ensure everything was up to par. He wanted to guarantee everything was going exactly to plan so the observations could be made with unwavering, pinpoint precision. The team decided to spend a few nights in Archangelsk before taking the week-long journey to Novaya Zemlya. And whilst they were there, they made final preparations, ate scrumptious food, and mixed with some of Russia's political elite. It was a charming few days for the team before the work would finally begin. But forever the inventor and businessman, Kristian believed that before the team disappeared over the horizon and into the depths of the Arctic for the second time, it would be good press to pose for a photograph, which could be used as promotional material for the forthcoming expedition. So, without wasting another minute, the team made their way to a local photography studio, groomed and ready to pose for the camera, in what they hoped would be an iconic photograph. For the photo, four of the men dressed in traditional Samoyed Arctic clothing as a bold statement to the harsh conditions they would imminently encounter. Whereas Kristian decided to sit in the centre of the group on a stone bench wearing his best suit, spectacles, bowtie, waxed moustache, and hat, like he was soon to attend a fine dinner party. Either way, it is a bold and captivating photograph that instills the bravery of such an expedition. And as each of the men had wished, the photograph did memorialise the occasion, and today, is one of the most famous pictures in aurora research history.

After Kristian deemed the photograph satisfactory, the party retired from the studio and made their way back into town to finalise any outstanding preparations. The mood was light-hearted as they exchanged words and quips about the trip to come. The team was excited, and *why would they not be?* They were set to embark on a second great expedition that they hoped would rock the world of science to its core, whilst simultaneously sating Kristian's thirst for understanding. As the team crossed the street a little way from the photography studio, still chatting through plans for the winter, a stray dog pulled itself from the mud at the side of the road and watched the men cross with interest. It was a rough dog that had evidently spent all its life on the streets as a stray. Its teeth were decaying and its lips blood-stained. It watched the party move and began to snarl at the group as they casually made their way past, perhaps infringing a little too much into what the dog believed to be its territory. It wasn't unusual to see stray dogs in these areas, but this one looked particularly ragged. Its fur matted and dirty and its eyes hard and bestial. The dog approached the party with calculated aggression and for some reason, became fixated on Kristian. One team member who was an experienced husky handler attempted to shoo the dog away, but it wasn't fazed by his efforts – it was transfixed and began to stalk the group. As they proceeded to stroll, the dog continued to follow, always growling and maintaining eye contact with Kristian. Again, they tried to shoo the dog away without success. As soon as the mutt had Kristian within its reach, it pounced, sank its yellow teeth into his calf whilst ripping through his favorite suit, and penetrated deep into the tender flesh of his leg. As Kristian yelped in pain and began to scuffle with the dog, another member of the group drew a gun and shot it dead to prevent it from attacking anyone else or causing further harm. But given its body language and unprompted aggression, the team became concerned that the beast may have been rabid, which as we can imagine, was not welcome news to Kristian. Soon afterwards a doctor was called to the scene. He cleaned the wound with a dosing of iodine in the hope of preventing infection, then wrapped the bite in lint. After a detailed inspection the doctor confirmed what Kristian did not want to hear: the dog had been rabid. Kristian was livid, for he knew at that moment his participation in the expedition was over. To add further distress to his situation, being the highly educated man he was, he knew the severe effects that rabies could have on the human body, and he was not willing to take any chances. He knew it could take more than four weeks for any symptoms to appear, and once they did, he was aware he may suffer from a series of convulsions, hydrophobia, and delirium before eventually meeting a rather unpleasant, undignified death. However, a rabies vaccine had been created and loosely tested by this point in time, but it was infamously difficult to procure, and there were no guarantees it would work due to the limited number of trials. With this rather distressing knowledge in hand, Kristian did what he always did: he got to work. He ordered his men to continue as normal in his absence whilst he planned to rest and recover in hospital. He figured he needed enough space and time to ensure he received the best care possible, alongside a strong dose of the rabies vaccination. Kristian was a resourceful and well-connected man, so he knew one way or another he would get his hands on the elusive vaccination. He also knew the expedition would still go ahead, and the observations would be made even if he played no part. Science waited for no one, including Kristian Olaf Bernhard Birkeland.

Shortly after Kristian made a reasonable recovery, he embarked upon a journey to Kaafjord, Finnmark, to pass his eyes over the aurora observation station he was supposed to have overseen. This station had been managed by Richard Krekling in his absence, who had done a superb job of managing the operations whilst Kristian was recovering in hospital. He'd ensured all equipment was sufficiently calibrated, and that each aurora display had been monitored and measured with the utmost accuracy. That winter, Richard had witnessed an impressive twenty-seven aurora displays. One of which, Kristian also viewed after he visited the station once he'd recovered from hospital. Kristian wrote of this display that:

> *The aurora of 24 November in particular was one of*
> *extreme beauty. It developed into an aurora corona,*
> *which lasted some minutes, then dissolved into a great*
> *number of intensely brilliant, red streams. These moved*
> *backwards and forwards across the heavens for some*
> *time, making the sky glow with red.*

But one point that stuck with Kristian after his limited time in the Arctic was not the number of observations the team had made, but the number of displays they had regrettably missed, despite their best intentions. The clouds and unpredictably aggressive weather of the Arctic meant many aurora displays were not seen, nor could they be accurately measured due to adverse conditions. Whilst the teams were in the Arctic, temperatures reached as low as -42c, and the winds proved so strong that making measurements often became nigh on impossible. To showcase some of the difficulties and dangers experienced by the teams over this second expedition, below are a few notes, letters, and extracts from diaries that were made during and after the expedition had finished.

The below is a letter written to Kristian from a team attempting to reach their observation station in Iceland, many days before their measurements and research had begun.

The station in Iceland was established much later than we had hoped. On the journey to Scotland we were delayed due to bad weather and again in the Faroes for another couple of days. By the time we reached the first stop in Iceland we were very late. We rushed around the Icelandic ports and saw the sun for the last time in a couple of months. More fog. More delays. On the last day of October a very heavy snowstorm began and we were forced to stay in Reykafjord for five days. It snowed non-stop and so thickly you could only see a couple of ships' lengths ahead. When we finally arrived at Captain Berg's whaling station the wind was so fierce that a portion of the roof was blown off even though it was covered in gravel and earth and tied down with steel wires. The Whaling Station is situated near the isolated promontory I have chosen for the measuring stations but the weather is creating great difficulties – overcast, deluges of snow and rain, high winds. It is the most remote place possible.

The next extract is a diary entry from the team situated in Novaya Zemlya, Russia. This entry documents the danger the team encountered once they ran out of fresh food and needed to head into the wilderness to hunt before they starved.

We left in a rowing boat and landed on the far side of a little river that could be easily waded. The boat was moored to the bank. Within hours and without warning a terrible storm broke out, with thunder and lightning which was very unusual in those parts. On returning to the boat several hours later the effect of the storm was frighteningly visible. The small stream had become a veritable foaming torrent and the entire tongue of land on which the boat had lain had been washed away. It was clear, however, that we must at all costs manage to get home. The fare was not first-class, it consisted of one dish – raw bird. With some old rope and driftwood we made a kind of raft and also found some boards that could be used as oars. It was an exceedingly poor vessel; even when we all three rowed with all our might it made only the slowest progress. When we got into the river current, we were carried rapidly out to sea and were soon several kilometers from shore. The worst of it was that the raft began to fall to pieces, so that one man had to hold it together with his feet and hands whilst the others rowed. After a hard struggle we reached an iceberg that was at least grounded and did not drift. Once more we took the oars and were fortunate enough to get into the counter-current, which carried us shoreward. Once we reached terra firma we saw how great the danger had been, for a fog as dense as a wall came pouring down from the north. If this had come a little sooner while we were rowing, it is highly probable we would have rowed in a circle whilst the current took us farther and farther there out.

The harshest of conditions were faced by the team in south Spitsbergen, Svalbard. They were isolated to such an extent that outside communication was impossible. Below is a tense extract from a letter written upon the team's eventual return home from the expedition.

It must be in great measure due to the tremendously varying conditions of weather that the immense loss of life on Spitsbergen is due. It is no exaggeration to say that all round and about our station is one great graveyard. It is for this reason that no one of late has ventured to winter in Spitsbergen; it is only during the last three or four years that it has been done once more, for the polar bear hunting. While we were building the instrument house a hurricane blew up so strongly that it was impossible to stand upright. It was a regular Spitsbergen storm in all its wildness and greatness. We were awakened by the roar and noise occasioned by wind, ice and rain. The wind varied incessantly; at one moment there was none, or a slight breeze, the next it was blowing the wildest hurricane. It was these fearful gusts of wind that were dangerous to anyone going out, for it was impossible to keep one's balance and gravel, snow and stones were whirled about. Although we managed to finish it, the hut was torn to pieces and a huge wooden panel hurled a hundred meters away. It will be easily understood that weather such as this places enormous difficulties in the way of observation.

Given the severity and terrible dangers documented in the extracts above, each of the teams managed to return home alive, much to

Kristian's relief. And with them they carried an abundance of fresh evidence that simply awaited scrutiny. Now, much to Kristian's delight, the analysis of the measurements and observations could begin. And he was as giddy as a child.

Kristian had long believed that if he could develop a good understanding of the auroras and their operation, he would be able to take this knowledge and recreate the lights in a controlled laboratory experiment, therefore proving his theories supreme. And now, with the new findings brought back by his loyal teams, he had the knowledge and expertise required. With the observations and measurements his teams had made on their latest expedition to the Arctic, Kristian knew he was on the precipice of scientific discovery. He also knew that once he'd proven his theories through scientific experimentation, his superiors and contemporaries would have no option but to bow to his research and acknowledge his discovery as being one of the greatest of the century. *And what better way could there be to prove his theory than by recreating the auroras himself in a fancy glass box?*

It was during these experiments that Kristian proved to the world that if he fired a cathode ray towards a magnetised brass ball with a phosphorescent surface, the ray would be pulled towards the magnetic poles of the ball. This interaction between the fired rays and the ball's magnetic force would then cause miniature reactions, which in turn, created glowing beams of light that surrounded the ball. In essence, Kristian had created his own Northern Lights in a large glass box in a dingy laboratory in rainy Oslo. To put it more simply: he had solved the riddle of the Northern Lights. And I'm sure if we could ask Kristian today if two dangerous expeditions to the Arctic, a fair amount of frostbite, the spending of his entire fortune, and a sever dousing of rabies were worth it, he would say, 'of course old chap, and I'd do it all again in a jiffy' – in a Norwegian accent, of course.

From then on, Kristian hypothesised that the electrons that reacted within the Earth's magnetic field must have been generated by sunspots on the solar surface. Somehow, these sunspots released electrons which later entangled themselves within the Earth's magnetic field and created the aurora. This theory also proved correct and secured Kristian a permanent position within the world's history books. If Jean-Jacques could see me typing these words, I have no doubt he'd be smiling in his grave whispering, 'I told you so'. But as great as Kristian's discoveries may have been, they were not fully appreciated. Unfortunately, ego can play a rather large role in the hindrance of scientific development. Many of Kristian's peers still held fast to their own ideas in the hope they would be the ones to solve the riddle. They hoped it would have been their names that filled the history books, not Kristian's, and that they would be the ones remembered for their contribution to human knowledge. But as science always does, as is its purpose, it pursued the truth and eventually proved Kristian correct. By which time the profoundness of his discovery had blown over and he was little congratulated for his achievement.

Kristian later released a book in 1908 named *The Norwegian Aurora Polaris Expedition of 1902 – 1903,* which greater detailed this theory alongside his supporting evidence. It was a publication that would change the world of science forever. Within his book, Kristian offered a diagram that helped to explain how the Sun's electrons travelled down the magnetic boundary of Earth. So, it's not surprising that today, this very diagram can be found on the back of a Norwegian 200 kroner banknote. As well as this, the banknote shows Kristian's face in a watermark next to the terrella diagram. And on the back of the note, only visible under ultraviolet light, is a picture of his rudimentary magnetosphere. The note is a true testament to the magnificence of Kristian's mind and discoveries.

After the release of Kristian's book, two other discoveries emerged that strengthened his theory on the Northern Lights. Firstly, in 1925, the Carnegie Institute of Science based in Washington D.C proved that the ionosphere, which is a vacuum-like region contained high up within Earth's atmosphere, began around fifty miles above the Earth's surface and continued to extend hundreds of miles higher into space. This is important because the ionosphere is an area where electrons can zip around, unimpeded, meaning the particles emitted from a solar flare can collide with the electrons in this region and emit the glow that is essentially an aurora. Further to this, the Norwegian mathematician and physicist, Carl Størmer, with his elfish ears and flared mustache, decided to measure the height of the auroras through a process of triangulation. Størmer had been an avid follower of Kristian's work and was quite fascinated by the aurora himself, where his main body of study had been the trajectory of particles and how this trajectory was influenced by Earth's magnetic sphere. His work was so successful that he laid the foundation for predicting particle movement around a sphere. He also published over 48 papers on the subject making him one of the world's authorities on particle and spherical interaction. Størmer's idea was to measure the aurora's height to see if this would somehow strengthen his research on particle movements. Measuring the height of the auroras proved to be no simple task, and throughout the process, Størmer took numerous measurements and photographs of the lights from 20 different observation points across Norway. What he concluded was that the Northern Lights typically shone around 100 kilometres above Earth's surface, whilst sometimes reaching as high as 1,000 kilometres into space. This discovery coincided perfectly with the Carnegie Institute of Science's findings on the ionosphere, showing that Kristian's theories on the lights where pretty much exact.

Størmer later published his findings on aurora height, which further cemented Kristian's discoveries as correct both within science and the public eye. These works inspiringly titled *From the Depths of Space to the Heart of the Atom* and *The Polar Aurora* were translated into five languages and remain to this day, a standard reference book on particle movement.

Now the riddle of the auroras has been solved thanks to Kristian and his team, I suppose the next question on our lips should be, *where do we go from here?* Because as you can see, this is not the end of our story. In answer to this, much has happened since Kristian decided to overdose on anti-paranoia drugs and end his own life, and many more people have dedicated their lives to further developing our understanding of the auroras. There is still much left to learn and many more stories to be told. But before we leave this chapter behind and take the next step on our journey, I would like to leave you with some words from Kristian Birkeland himself, as he talks about his terrella experiment and the Universe as he knows it. I hope you enjoy them as much as I did:

> *To understand the distances that I have captured in this vessel, imagine that our sun is a grain of sand a millimeter in diameter. In that case, the Earth would be an invisible speck of dust ten centimeters away. And the next nearest star, Alpha Centauri, would be twenty kilometers away. It is in this vast, infinite space that the genesis of all celestial bodies is to be found. All matter that we see, be it our own bodies, our Earth, other planets, the sun, our solar system, and other solar systems, all matter is composed of flying atoms that are continuously ejected from our sun and other suns by electrical forces and that condense and form particles. And these in turn condense to form large spheres, ultimately planets and all thereupon. It follows from this*

that everything that is matter, all living beings in the universe, is linked, one to the other.

It seems to me to be true, from what I have already said, that new worlds emerge in space more frequently than human beings are born on Earth. Each world has its éclair de nuit, its flash of lightning in the dark, the struggle of intelligent beings with their own thoughts and their discoveries to banish ignorance. Such moments of illumination disappear without trace, and it also follows that such worlds must die more frequently than human beings die upon Earth or, more accurately, they are born and die in such number that surpasses our imagination.

STANDING ON THE SHOULDERS OF GIANTS

'If I have seen further, it is by standing on the shoulders of giants.'

Isaac Newton

Since Kristian's death, research into the auroras has only gained momentum, and many organisations and bodies have, and are, adding to our knowledge each day. One such body is the Rutherford Appleton Laboratory located in Chilton, Oxfordshire, England, which employs 1,200 budding individuals who support the tireless work of over 10,000 scientists.

Of all the scientific research this laboratory undertakes, it is the Double Star Polar project that investigates the interaction between Earth and the Sun that is the most important when it comes to understanding the auroras. One of the most significant missions undertaken by this programme is the Cluster II space mission. This mission, managed by the European Space Agency (ESA) with the assistance of NASA, aims to study the Earth's magnetic field under the influence of the Sun, and how the Sun's varying activity causes the magnetic field to react. In order to do this the ESA launched four identical spacecrafts which travel in a tetrahedron formation (a three-dimensional triangle) throughout space. These spacecrafts were launched from Kazakhstan in the year 2000 after the initial mission, Cluster, botched in 1996 when all four spacecraft disappointingly failed to reach orbit. If we wanted to be simplistic, we could say the four spacecrafts of Cluster II have one overriding objective: to collect three-dimensional information on how the Sun's activity interacts with Earth. And as you have probably guessed by now, it is this activity that creates the auroras. During their quest for three-dimensional information the Cluster II spacecraft rarely break their tetrahedron formation. But even though this structure is maintained, the spacecrafts are sometimes as far as 10,000 km apart, which is the equivalent of one spacecraft sitting in London, UK, and another residing in Adelaide, Australia. Once the three-dimensional information has been collected it is beamed back to Earth where the scientists analyse and interpret the data, and thus, achieve a greater understanding of the Sun's activity, the Earth's response to that activity, and the cause and effect of the auroras. Over the years the Cluster II mission has been operating, there have been many notable milestones and discoveries, such as: *How Cluster II Helped to Protect Astronauts and Satellites from Killer Electrons; Space is Fizzy; How Space Tsunamis are Formed;* and *Solar Wind is Swirly*. I'm not sure of the cost of the Cluster II mission, but however much it may be, I'm sure it's money well spent when we can attest to such profound discoveries as *Solar Wind is Swirly*. Nevertheless, after 16 years of robust research and information collation, this scientific endeavor regrettably came to an end in December 2016.

Cluster II is not the only astronomical research mission, however. As high within the outer edges of Earth's atmosphere at 250 miles above sea-level floats an engineering wonder the same size as a six-bedroom house: The International Space Station (ISS). The ISS undertakes a spectrum of research that spans the biological and physical sciences of Earth. All of which is undertaken for the benefit of each of us as we plod along day after day, living out our everyday lives. It's nice to think that as you sit on the sofa, drink beer in the pub with your friends, or walk your dog through the park, there are a series of scientists floating around in space undertaking research for your personal benefit. The ISS has been continually occupied since November 2000, and at the time of writing, 222 people from 18 countries have visited. What's more, the ship travels at such speed that it circles Earth once every 90 minutes – this equates to five miles per second. Imagine a six-bedroom house travelling at five miles per second! It also glides across the night sky as the second brightest object visible, after the Moon. So next time you look up to the sky on a clear night, when the ISS is in your vicinity, look for the second brightest object overhead (I'm hoping you can figure out which one the Moon is) and give the inquisitive scientists aboard a silent thank you. The reason I mentioned the ISS is that it sits in a prime location to view the auroras. From their vantage point they can witness the auroras from an angle never before seen. From their seat in space, they can (and have) watched the long, stringy body of the auroras dance on top of Earth over both the northern and southern magnetic poles. It's an unbelievable sight to behold, or, in the words of astronaut Don Pettie it is '…like being shrunk down and put inside of a neon sign.'

As we carry on naming scientific research centres and vessels, sitting within Uppsala, northern Sweden, close to the point where the Finnish, Swedish, and Norwegian boarders collide, is a governmental research centre named The Swedish Institute of Space Physics (SISP). The centre's task is to undertake basic research, education, and associated observatory activities within space physics, space technology, and atmospheric physics. Alongside the SISP's research into the auroras it is also undertaking missions to Mars, Saturn, Mercury, and the Moon, and will soon be venturing towards the icy moons of Jupiter. It sounds like a truly fascinating place to work. The institute was originally set-up in 1957 under a different name, when its main objective was to monitor aurora activity as it sat in a prime location to observe and study the lights. This activity was assessed by sending rockets into the night sky when the auroras appeared. The rockets would be tasked with collecting particles from the active aurora, which would later be tested and analysed upon their return. I struggle to think of anything more amazing than sending a rocket into the heart of the aurora to collect particles from the Northern Lights. It sounds like the stuff of fairytales. But as technology developed these rockets were used less frequently, until satellite-based research completely took over. The first satellite launched by the SISP was named, 'Viking'. An incredibly apt name as we have seen in an earlier chapter, considering Sweden's history, and the past mythology that surrounded the auroras. Again, the story of the auroras is tainted with irony, as the Viking satellite ventures high into the atmosphere to study and collect particles from what was once believed to have been the

Bifröst. If the ancient myths are still to be believed today, then this could be perceived as the humans of Earth developing technology so powerful, it is bringing them within a cat's whisker of their gods: an incredibly inspiring thought.

Therefore, with many scientific endeavors such as Cluster II, the ISS, and SISP happening across the globe, it is with a certain amount of confidence we can now say we understand the auroras, how they are formed, and what they are. Science has shown us that they are almost certainly not caused by dancing fish or football playing walruses. They are not formed by a giant fox waving its tail, or a trapped swan frantically flapping its wings. And we are almost certain they have never had sex with anyone. Because as with all things, there is a more logical explanation.

THE SCIENCE BIT

'Physics is like sex: sure, it may give you some practical results, but that's not why we do it.'

Richard Feynman

It all starts with a solar flare. The outer layer of the Sun, known as the corona, contains a series of sunspots. Close to these sunspots are intense magnetic fields which can penetrate the photosphere (the Sun's outer shell) and link the corona to the Sun's interior. Once a connection between the corona and the Sun's interior has been established, the corona can climb in temperature to a dazzling 1.1 million degrees Celsius. When the corona reaches such a temperature, the Sun's gravity struggles to hold onto the rapidly moving particles on the surface – because as we all learnt in school, the hotter the temperature, the more violently particles react. As the particles increase in temperature and progressively amplify their rate of vibration, their outward force becomes greater than the gravity's inward force causing them to explode outward and away from the Sun. This outward burst is known as a solar flare. The power of a solar flare is astonishing, roughly the equivalent of 160,000,000,000 megatons of TNT. As a basis for comparison, the most powerful explosion ever created by humans was that of the Tsar hydrogen bomb: a nuclear warhead designed and detonated by the Russians in 1961, which released the equivalent of 58 megatons of TNT – a mere drop in the ocean compared with a solar flare. It is also known that the Sun's activity incrementally climbs over an 11-year period known as a 'solar cycle'. This means that radiation levels and solar flare activity will vary depending on the time of the cycle they fall within. This 11-year cycle also means that the auroras slowly strengthen and become more active over the cycle period. The first year of this cycle is known as 'solar minimum', where solar flare and aurora activity is often small or negligible. As you would have guessed, the last year of the cycle is aptly named 'solar maximum', where solar flare and aurora activity is often vigorous and sprightly. So, as you can see, aurora activity is directly linked to solar activity. What's more, in a rather perplexing sort of way, is that once the Sun has completed its 11-year cycle its magnetic poles will flip, meaning 'north' becomes 'south' and vice versa. This unexpected upending of the poles bewildered solar physicists for many years, but now we know why it occurs. The reason the poles of the Sun flip is because sunspots are an intense hive of magnetic energy, and as the sunspots become more widespread over the solar cycle period, the cumulative magnetic energy they contain as a grouping grows to staggering heights. Then, as they drift across the Sun's surface, the positively charged sunspots are attracted towards the negatively charged pole, and contrariwise. It gets to such a point that eventually the magnetic influence of the increasing sunspots becomes greater than the magnetic power of the pole that attracted them. This inundates the pole causing an upturn in magnetism, which forces the Sun's magnetic field to flip.

Once the solar flare has detonated, the mass ejection released from the Sun travels through space as solar wind. As we can imagine, due to the enormous power of a solar flare, solar wind can travel extremely quickly – sometimes up to 500 miles (800 kilometers) per second. This wind contains a soup of electrons, ions, and atoms, which due to their excessive speed usually reach Earth within one or two days. If solar activity is particularly violent then bubbles of gas and magnetic field can explode from the Sun, releasing significant amounts of plasma, known as coronal mass ejections (CMEs) or solar storms. These CMEs snap outwards with such force that their contents (around a billion tons of matter) are the equivalent of the total mass of Mt. Everest. This means when a CME explodes from the Sun, we have a soup of particles the equivalent to that of the highest mountain on our planet, flying towards us at millions of miles per hour, aiming for a head on collision with our atmosphere. As you would imagine, these solar storms or CMEs have a stronger effect on Earth's magnetic field than standard solar wind, and therefore, a more active and prominent aurora is formed.

It's times like this we should be thankful for our ever-active magnetic field, because without it, the solar radiation and plasma that's released from the Sun's giant solar flares would collide with Earth and quite frankly, kill us all. The magnetic field protects us from the Sun's harmful expulsions, and in turn, transforms them into something beautiful: the auroras.

When solar wind approaches Earth it makes contact with the magnetosphere – the section of space which the Earth's magnetic field occupies. As the particles enter the magnetosphere they are pulled towards the most magnetic points of Earth's magnetic field: the poles. It's at these points that the particles make contact with the magnetic field, which captures some of the particles and sends them spiraling down through Earth's atmosphere. Once the particles that have been captured by the magnetic field are pulled further towards Earth, they collide with the gases that are contained within the atmosphere, such as nitrogen and oxygen. And due to some of the particles contained within Earth's atmosphere being reactive, once solar wind collides with these particles they light up and create an aurora. If these reactions take place around the northern magnetic pole they are named, aurora borealis; if they occur around the southern magnetic pole they are called, aurora australis.

A further note made on the auroras is that regardless of which pole they appear, their performance is rarely a one-night show. The reason for this is that a solar flare releases a truly profound number of particles, which can often collide with the magnetic field for several days. So, a good night's aurora viewing is often immediately followed by another – cloud-cover permitting. It has also been found that solar flare activity can continue to churn on the same spot of the Sun for months on end. *What does this mean?* Well, considering the Sun rotates on an axis, it means that once the Sun has undertaken its usual rotation, the solar flare producing zone will be back facing Earth again, spewing particles towards us that will ineluctably react with our magnetic field. So, if you miss a great aurora performance for one reason or another, sometimes you can head back out again 27 days later and try to catch it again – kind of like your favorite TV show on repeat.

To add to this point, scientists have discovered that the auroras of both poles are not completely random in their movements and pattern. They pretty much always follow a certain, predictable routine known as a 'substorm'. The first stage of a substorm is that the light will appear very dimly on the northern/southern horizon (depending on which hemisphere you're in) usually within the late evening of a crisp winter's night. At first the light will have no real structure nor colour to define it by. It will appear hazy and diffused as if it's just a glowing bundle of luminous, colourless clouds. Then, as the lights start to amalgamate, they will usually form an arc shape that bends like an archer's bow across the sky. This arc shape is also pretty much always pale green to begin with. Sometimes the arc will remain green and other times it will change colour to either red, orange, blue or a spectrum of other colours all displayed at once. But regardless of colour and intensity, the light will soon start to stretch outwards until it reaches towards the eastern and western horizons like a giant rainbow. To a curious spectator, this giant rainbow looks like it stretches towards the surface of Earth on either end, but this is just an illusion. It's actually the light following the curvature of Earth as the particles of the solar flare trace along the lines of the magnetic field. In-fact, one single beam of aurora light can span over 1,000 miles in length, which is the equivalent distance from London, UK to Lisbon, Portugal – hence why the curvature of some auroras is so evident. Auroras are also a lot taller than most people think. An average beam can be 100 miles tall, whereas a powerful, thick aurora beam can stretch up to 500 miles from base to tip. Yet given their potentially grand height and length, their width can appear somewhat lacking in comparison at around one mile, give or take. So, with such a gargantuan aurora shining in the sky you will not be surprised to learn that the largest of auroras can be seen from over 1,000 miles away, unaided. As the aurora grows in intensity it progressively becomes more active and will begin to develop a series of folds, which look like the effect you get from an open, hanging curtain. These folds appear stationary at first, but they will soon begin to ripple from side-to-side as the aurora grows larger and heads towards the southern/northern horizon. The direction of the folds' movement can change as it sees fit, so that instead of just rippling along a single plain, it can move in all possible directions including up and down. This movement creates the effect that the aurora is constantly climbing upon itself as if it's trying to reach higher, before collapsing downwards as if tumbling down a series of stairs and back into itself. As the ripples move the aurora becomes somewhat unpredictable, like a wildfire in the sky, enhancing and retreating wherever it can. If you are to watch an aurora closely as it ripples, moves, and dances across the sky, you may also witness a point within the light that appears to show beams permeating the sky in all directions, like rays of the Sun. As spectacular as this appears to be, I'm afraid to tell you that you're being tricked by the aurora, and what you're seeing is an illusion. The point of light that's conjuring the trickery is aptly named the Corona and is effectively caused by your relative position to the light. You see, unfortunately there is no miniature explosion of light contained within an aurora, as all aurora rays are pretty much vertical.

Once the particles of a solar flare have finished their collision with the magnetic field and its contents, the auroras will begin to fade. They will diffuse into a plethora of directions like a fading firework, until eventually, there is nothing more than an indistinguishable haze that fades into darkness, leaving nothing behind.

Now you may be surprised to learn that over the years, science has discovered that on occasion, the Earth's magnetic field has also been prone to inverting upon itself – just as the Sun's does, as we discovered earlier. The last time this happened is believed to have been roughly 800,000 years ago, so it is not a major cause for concern. But scientists have not yet been able to pinpoint the reason why this inversion takes place, so are in no position to preempt it should it occur again. This means that a polar inversion could occur at any moment. But an inversion is by no means instantaneous; it's a laboriously slow process. Some estimates have claimed that a full polar inversion could take as long as 5,000 years. Thus, the Earth's magnetic poles could in theory be placed anywhere on Earth during this transition from north to south, and vice versa. This slow transition would therefore allow the auroras to appear in contrasting locations to where we see them today. Imagine if the magnetic North Pole was positioned above Wigan, and this is where the auroras shone brightest. Tourists would flock to Wigan each winter to see its natural wonders laid bare. Fortunately for us, this dystopian existence doesn't yet exist.

Most aurora activity takes place within an area known as the 'aurora zone', which is a ring-shaped region with an approximate radius of 2,500 km that sits over Earth's magnetic poles like a giant halo. This zone is highly reactive to the Sun's activity and when solar flares and CME's are particularly strong, they can collide with this zone and cause it to stretch. This is what happens when the auroras can be seen at the lower latitudes. The aurora zone is literally stretching under the force of the collisions from the Sun's ejections. This aurora zone was first discovered in the 1970s when it was thought to have only appeared at night, when there was sufficient darkness and solar activity for it to be seen. But due to scientific advancement it has been discovered that the aurora zone is constantly displaying aurora activity; every day, every night, seven days a week, 365 days a year. The reason we don't notice is because our eyes are not sensitive enough to see all the light. But if you were to monitor the aurora zone (as many people have) you would find that with the right equipment the auroras are constantly dancing. What an astonishing thought! It's exhilarating to think that the auroras are continually shining above our heads, all day every day. As a matter of fact, some scientists have coined the aurora zone the 'window to the universe', where we can constantly witness what's happening outside our own tiny world by monitoring aurora and solar activity. But not all aurora activity takes place within this 'window to the universe'. If a solar flare is particularly strong and releases an abundance of solar wind, then more particles will collide with Earth's magnetic field whilst travelling at greater speed. This means the auroras will travel further into the Earth's atmosphere and form greater, brighter reactions when they encounter particles. It's in episodes such as this when the aurora borealis can be seen further south, and the aurora australis further north. And if the colliding of particles is abundant and powerful enough, it is possible to witness the auroras closer to the equatorial regions. There's a very high possibility this is exactly what happened when Aristotle witnessed and wrote about the lights in Greece, 344BC.

If you wanted to tread in the 'footsteps' of Aristotle and go to witness the auroras in action, you would be best advised to head for northern Canada or Alaska, USA. Heading as deep as you can into the Arctic Circle whilst maintaining your position on terra firma. The reason for this is because the auroras appear here more frequently than anywhere else on Earth, which is thanks to the slanted positioning of Earth's magnetic field. You see, the magnetic field doesn't sit perfectly over Earth. It's crooked and tilts to one side on a jaunty angle, like a 1950s gangster's trilby. This is because Earth's geographic and magnetic poles are to be found in different locations. If you followed a compass needle as far north as you could, you would not end up at the most northerly point on Earth (the geographic North Pole, also known as 'true north') you would find your way to the magnetic North Pole, which is roughly 800 miles further south-west due to Earth's slanting magnetic field. This slanting means the Canadian mainland and Alaska are closer to the magnetic North Pole than any other landmass on Earth. This is why it is the best place to witness an aurora show. The lights appear here more frequently and more brilliantly than anywhere else around. If you headed incredibly far north to the likes of Chalkyitsik, Canada, you would be almost guaranteed to see the lights every night throughout winter. Not only that, but as you stood there, you would be able to see all the light's colours; the reds, blues, greens, and purples dancing alone and intermingling in a Technicolor display. One of the most fascinating parts of an aurora display is the beautiful assortment of colours they showcase. Sometimes they are bold, solid, independent colours. Yet on occasion they are a mixture of a variety of vibrant and opaque shades. And all of this is thanks to the invisible bubble that surrounds our planet. A protective casing that enables Earth to harness life: the atmosphere.

THE ATMOSPHERE

'When the earth came alive it began constructing its own membrane, for the general purpose of editing the sun.'

Lewis Thomas

The term atmosphere refers to the layer of gases that surround Earth, which are commonly known as air. The atmosphere has a general make-up of Nitrogen (78.09%), Oxygen (20.95%), and Argon (0.039%), with the remainder consisting of small amounts of other gases – including those you blame on the dog. These are generic statistics for the whole of the atmosphere so will not be applicable to all layers, because each time you change altitude the gaseous make-up alters. But they are sufficient for our purposes. This is why scientifically the atmosphere has been broken down into five main layers:

Exosphere (440 – 6,200 miles)
Thermosphere (50 – 440 miles)
Mesosphere (31 – 50 miles)
Stratosphere (7 – 31 miles)
Troposphere (0 – 7 miles)

Let's start at the top and work our way down.

The Exosphere is the outermost layer of the Earth's atmosphere. Contrary to the above stated heights, the Exosphere actually has no definite ending point. It simply diffuses into space, meaning there is no definitive line. At this point of diffusion (which is usually coined as 6,200 miles above Earth's surface) the Exosphere merges with solar wind making it the starting point of the auroras. Yet, if there is any aurora activity at this level it cannot be witnessed by the human eye, as the particles at this height are sometimes hundreds of kilometres apart – meaning any aurora activity that is created is so sporadic and dim it's untraceable. The particles of the Exosphere are so distant from one another that they often drift off into space without telling anyone, never to be heard of again. This is why sometimes the Exosphere is not considered a worthy part of the atmosphere and in some schools its existence is ignored completely.

If we are to travel down further into the atmosphere and out of the Exosphere, we will encounter the upper limit of the Thermosphere at the altitude of 440 miles. This is the second highest layer of the Earth's atmosphere, and this layer is especially susceptible to solar activity. This means its height can vary considerably depending on if the Sun is highly active or dormant. The Thermosphere is incredibly hot, sometimes reaching temperatures of up to 2,000 Celsius. But as the particles are so far apart (usually around 1km) you and I could stand in the middle of the Thermosphere and not get burnt, as there would not be sufficient molecule contact with our skin in order to transfer enough heat into us for it to burn. This region of the atmosphere is also completely water and cloud free and is the layer within which the International Space Station can be found. It could also be said that as we are starting to move lower and the particles marginally closer together, aurora activity is more likely to appear in this region. But it's slightly different in colour and activity level than it would be within the Exosphere, due to the density and particle types present at this altitude.

Next, we will take another step downwards and enter the Mesosphere, which is classed as the coldest place on Earth with an average temperature of -85 Celsius. This is a stupendous contrast to the temperature of the Thermosphere above, which highlights the mass differentiation of the atmosphere's layers and their aurora producing capabilities. This region is practically a purgatory between Earth and space. It is too high to be accessed by jet powered aircraft, yet too low to permit orbital aircraft. It's home to a few sparse noctilucent clouds that are formed from instantly frozen water vapour that dares to venture so high, and it destroys most meteors flying through space that dare to venture so low. It's a hostile limbo between life on Earth's surface and the black void of space above.

Below the Mesosphere is the Stratosphere, which is the second lowest layer of the atmosphere. This is the part of the atmosphere that contains the famous (or infamous, depending on how you look at it) ozone layer. There's a much higher concentration of gas at this level, which allows it to absorb a fair amount of the Sun's ultraviolet radiation.

Finally, we enter the Troposphere, the area in which you and I are now sat. This layer reaches out to an altitude of seven miles and contains 80% of the atmosphere's total mass, making it incredibly good for many things that you and I like to do, such as breathing. Due to the abundance of particles and molecules at this level, it's the region where pretty much all life on Earth exists. Clouds float here, airplanes fly, animals live and breed, and you and I can talk with our friends about that good book we're reading – I'm very aware it may not be this one. Overall, it's an incredibly hospitable layer.

But *what does all this have to do with the colour of the auroras?* Well, I'm glad you asked. The colour of the auroras is dictated by a handful of variables, yet the simplest way to think about this is to compare the auroras and Earth to a neon street sign filled with gas. When somebody wants to light a gas filled street sign, they turn on the electricity, allow the electrons to collide with the gas, and the colour generated by the reaction depends on the type of gas contained within the sign. The auroras are effectively the same. Think of Earth as a big, round, neon, gas filled street sign, floating idly in the Milky Way. The solar wind released by the Sun acts as the electricity, entering Earth's atmosphere and colliding with the gaseous particles contained within whatever layer of the atmosphere they hit. As a result, the colour of the auroras depends on which gaseous particles the solar wind collides with, and which gaseous particles the wind collides with depends on which layer of the atmosphere the solar wind engages with. This simple action explains why the colours of the auroras you see are directly linked to the altitude at which the collisions and reactions between the solar wind and Earth's particles takes place. Because at the different altitudes there are varying mixtures and densities of gas. Unfortunately, however, it is not quite this simple, as there are other contributing factors to the colours of the auroras. Things such as the amount of energy Earth's particles receive from the solar wind, and how the wavelengths of light blend with each other. These things can drastically alter the colours that the auroras appear to a humble spectator on Earth. But the simple and easy to remember rule of altitude, and the gases contained at those altitudes, is usually sufficiently accurate, and is widely adopted as a method of explanation.

If we were to place the colours of the auroras in order, from high to low altitude collisions, it would look something like this:

Red (transparent)
Green
Red (blood)
Yellow, Pink, Orange
Blue

The reason any of these colours are possible in the first place is attributed to the Sun, as sunlight is white, and therefore, contains all visible colours. As the solar wind collides with Earth's particles it causes the electrons contained within them to become excited. These electrons then vibrate for a short while before returning to their normal state. And it's once these electrons return to their normal state that they emit their colourful glow, which forms the auroras. In simple terms, if these collisions take part high-up within the ionosphere, around a 190+ mile altitude, the auroras will appear red due to the main atoms present at this altitude being oxygen. Although the main atoms present at this altitude are oxygen, there is on the whole a lack of any particles present in this level of the Thermosphere when compared to lower altitudes. This is why the red auroras that are formed at the highest altitudes will appear highly transparent to the human eye.

As for green auroras, their forming collisions take part at an altitude of around 75 – 110 miles (still within the depths of the Thermosphere) and are also formed by vibrating oxygen atoms. The reason the colour differs although the reaction is caused by the same atom is because at this altitude the oxygen atoms are present in a higher density. This results in a greater frequency of collisions, which releases colours at a different wavelength causing the auroras to appear green as opposed to red. The human eye is also more receptive to the colour green than red. This increased receptivity, alongside the greater frequency of collisions, is what makes the green auroras the most common of all auroras. But just to make matters slightly more complex, if the collisions take part somewhere between the 'green altitude' and 'red altitude' areas, these colours can mix and form pink or yellow auroras. This is why on occasion green auroras can be witnessed with a pink or yellow hue – because the edge of the aurora is effectively breaching a colour boundary.

Below green, red auroras can appear again. This is due to a change in gas mixture at this altitude, as at this lower altitude, the quantity of nitrogen atoms significantly increases. This means the solar wind collides with more nitrogen atoms at this altitude instead of the available oxygen atoms, and a powerful red colour is emitted as a result. The red auroras of this altitude are darker and more authoritative than those formed at higher altitudes within the ionosphere as there is a greater density of atoms to collide with. The red auroras formed by these nitrogen atoms are sometimes described as 'blood-red', as opposed to the higher altitude auroras which are a lighter, subtler, more transparent red.

This process continues as we progress past the yellow, pink, and orange auroras until we hit the colour blue, which sits at the bottom of the visible colour scale. Blue auroras are caused by nitrogen atoms at an altitude of up to 60 miles and are somewhat rare and infrequently appear alone. They are mainly seen as a skirting on the lower edge of a green or yellow aurora, which is formed by the little solar wind that makes it to this lower altitude without happening upon any collisions along the way. A blue aurora skirting is only usually seen during periods of intense solar activity, when the auroras are brighter, larger, and more active than usual – predominantly towards the end of the 11-year solar cycle.

It's once we reach this low altitude that the visible auroras begin to disappear. Not all is lost, however. As with a little technology in the form of a highly sensitive camera lens the auroras continue to shine. Although the human eye cannot see the auroras that are formed at these lower altitudes, due to the light being emitted at a wavelength that is untraceable to us, a high-quality camera lens can. If you were to record the auroras in action with such a lens, on occasion you would see ultraviolet and infrared auroras shining at very low altitudes. Sometimes even within the Troposphere you and I now occupy. So if you were to wrap up warm, head to the Arctic circle, and watch the auroras shimmy above; with sufficient technology you could possibly see the 'invisible' auroras dancing around you, zipping underneath your arms, over your ears, through your legs and in-between your fingers like a giant, glowing, interstellar worm. It is thoughts such as this that humble my soul. To think that there are things happening in this world that our senses can simply not digest. That sometimes at night, you and I may be walking amongst the aurora, completely unaware of our connection with the Sun. In reality, there is much in this world we cannot see, hear, or smell due to our biological limitations. And that makes Earth all the more magical.

Even though the dancing of the auroras appears pure, subtle, and delicate, sometimes the atmosphere and magnetic field of Earth must endure severe bombardments of energy that result in extreme aurora activity. These instances of high energy activity can be a major cause for concern amongst scientists and civilians alike –

no matter how pretty the energy looks once it manifests itself as an aurora. One such example of this occurred in June 2016 when the GRAPES-3 muon telescope in India, which is specially designed to detect high energy interactions, spotted a giant cluster of galactic cosmic rays (GCRs) travelling our way. These rays originated outside our solar system and had unnervingly combined themselves with a large coronal mass ejection from the Sun, and were heading for our planet. The GRAPES-3 telescope watched in anticipation as a head-on collision became imminent. The storm, which quickly rushed upon us, raged with such strength and ferocity that it was rated as G4, or 'severe', by the National Oceanic and Atmospheric Administration (NOAA). As it raced closer and closer the scientists watched and waited, with bated breath. Minutes later the collision occurred. Powerful auroras were emitted around the world. The auroras of north and south shone abnormally bright allowing them to be witnessed in places more unusual. It was a superb display, a testament to our Sun's power and beauty. But this is not all that happened. The GCR's and CME impacted with Earth with such brute force that our magnetic field cracked. Just take a moment to digest that thought: the magnetic field of Earth cracked under the pressure of the collision. As we can imagine, an act such as this caused more than a little panic. *Had the magnetic field fractured before?* Possibly. But we had never witnessed it. The magnetic field had also been magnificently compressed under the looming force of the storm. Usually, the magnetosphere spans a radius of 11 times that of Earth, but for a few tense hours, the magnetosphere's radius was reduced to a mere four times that of our planet.

Fortunately for us the damage caused by the storm only lasted a few hours. The magnetic field repaired itself and quickly expanded back to its usual size. Nevertheless, it was an exciting and tense few hours for our planet that the general public (more than likely including yourself) lived through without noticing a thing.

As we saw earlier, the solar cycle that naturally occurs causes the strength of the auroras to climb over an 11-year period. This cycle is highly predictable and usually operates exactly as expected. But sometimes the cycle experiences a blip, of sorts, and unusually quiet or worryingly aggressive periods of solar flare activity occur. It wouldn't be of much interest if I started to write about an unusually quiet period of solar activity, explaining and naming the years where very little happened. The times when the nights were darker than usual, and nobody saw anything out of the ordinary. That would be dreary to read and rather tedious to write. But I'm sure you'd agree, it would be rather fun if we looked at the overly active periods of solar activity. The real big blips that rarely occur. In-fact, *wouldn't it be interesting before we conclude our story if we took a very brief yet entertaining look at the most powerful geomagnetic storm in history?* The time on Earth when the auroras shone the brightest ever recorded. *And what were the consequences of such an event?* Well, it's time to take a trip back to the 19th century. To be exact, it's time to travel back to 2nd September 1859.

THE GREAT GEOMAGNETIC STORM

'Life isn't about waiting for the storm to pass...It's about learning to dance in the rain.'

Vivian Greene

The solar storm of 1859 nicknamed 'the great geomagnetic storm', is the largest, most spectacular solar storm ever recorded. The magnificence of this storm created the most powerful, brightly lit auroras ever documented on this planet. The storm occurred during a period known as solar cycle ten, which was the tenth solar cycle witnessed since records began in 1755. During this cycle the amateur astronomer Richard Christopher Carrington became the first ever person to witness a solar flare in action, and in consequence, prove their existence to science. However, little did the world know, nor Carrington for that matter, that the solar flare he had just witnessed through his long, peering telescope, was to become the largest solar flare in chronicled history.

It all began on 1st September 1859 when Carrington witnessed an unusually bright, white light, explode from the Sun's surface. This white light was the releasing of a solar flare that would travel through space for 17.6 hours before making contact with Earth's magnetic field. Upon contact, the velocity and density of the charged particles contained within the solar wind was so great that there were numerous, gigantic reactions with the gaseous particles of Earth. The auroras these reactions created were seen on both sides of the planet. The number of particles and energy contained within the solar wind was so magnificent, it collided with both magnetic poles at once, creating simultaneous aurora borealis and aurora australis. These exceptionally bright and magnetically disruptive auroras were recorded across the globe from Europe, USA, Australia, Canada, Japan, Senegal, Mauritania, Liberia, Mexico, Hawaii, and even Cuba. To place this into perspective, Cuba sits a mere 1,600 miles from the equator, meaning if the skies were clear and perpetual night had covered Earth, the auroras would have been seen around the entire planet, in every sky, no matter where you were. All you would have to do is look up.

The glow the auroras emitted was so bright that the gold miners of the Rocky Mountains were awoken by it, believed it was morning and began preparing breakfast. Telegraph systems all over Europe and the USA threw sparks and failed due to the gargantuan magnetic influx of solar wind. As if to cause further confusion, many telegraph systems magically rebooted even though their power supply had been destroyed. In a moment of disbelief, the telegraph systems that had been rebooted were now working from the power of the auroras. They were operating via a geomagnetically induced current that was being issued by the Earth's electromagnetic field. The *Boston Traveller* ran an article on this unbelievable event, where the periodical transcribed part of a telegraph message that had been transmitted from Boston to Portland through the power of the auroras.

Boston operator (to Portland operator): "Please cut off your battery [power source] entirely for fifteen minutes."
Portland operator: "Will do so. It is now disconnected."
Boston operator: "Mine is disconnected, and we are working with the auroral current. How do you receive my writing?"
Portland operator: "Better than with our batteries on – current comes and goes gradually."
Boston operator: "My current is very strong at times, and we can work better without the batteries, as the aurora seems to neutralize and augment our batteries alternately, making currents too strong at times for relay magnets. Suppose we work without batteries while we are affected by this trouble?"
Portland operator: "Very well. Shall I go ahead with business?"
Boston operator: "Yes, go ahead."

In total, it was reported this conversation ran for a few hours via the energy of the auroras before power failed and the batteries had to be reinstated.

The aurora event was so prolific it was covered in multiple publications, with some of the articles written about the night proving quite amusing:

A brilliant display of Northern Lights was witnessed from 8 o'clock to half-past 9 last night. The glare in the northern sky, pervious to defining itself into the well-known features of the Aurora Borealis was sufficiently vivid to call out some of the fire companies.

– **The Evening Star, Washington DC, 1859**

...There was another display of the Aurora last night so brilliant that at about one o'clock ordinary print could be read by light...

– **The New York Times, 1859**

...Singular as it may appear, a gentleman actually killed three birds with a gun yesterday morning about 1 o'clock, a circumstance which perhaps never had its like before. The birds were killed while the beautiful aurora borealis was at its height, and being a very early species – larks – were, no doubt, deceived by the bright appearance of everything, and came forth innocently, supposing it was day...

– **New Orleans Daily Picayune, September 9, 1859**

Those who happened to be out late on Thursday night had an opportunity of witnessing another magnificent display of the auroral lights. The phenomenon was very similar to the display on Sunday night, though at times the light was, if possible, more brilliant, and the prismatic hues more varied and gorgeous. The light appeared to cover the whole firmament, apparently like a luminous cloud, through which the stars of the larger magnitude indistinctly shone. The light was greater than that of the moon at its full, but had an indescribable softness and delicacy that seemed to envelop everything upon which it rested. Between 12 and 1 o'clock, when the display was at its full brilliancy, the quiet streets of the city resting under this strange light, presented a beautiful as well as singular appearance.

– **Baltimore American and Commercial Advertiser, September 3, 1859**

One such chap named C. F. Herbert wrote an article for *The Daily News* in Perth, Australia, telling his story of the night and how his witnessing of the aurora australis unfolded. He details how the people of the local area reacted in contrasting ways to the lights, either believing them to be a form of divine intervention or an ominous warning from heaven. The article was submitted to the newspaper upon request in 1909, 50 years after the event, as a means of recalling the great night for the new generation who were born after its passing:

(To the Editor.)

Having seen the phenomena on 25th last, a short recital of the great aurora of fifty years ago might be interesting. I was gold-digging at Rokewood, about four miles from Rokewood township (Victoria). Myself and two mates looking out of the tent saw a great reflection in the southern heavens at about 7 o'clock p.m., and in about half an hour a scene of almost unspeakable beauty presented itself, lights of every imaginable color were issuing from the southern heavens, one color fading away only to give place to another if possible more beautiful than the last, the streams mounting to the

*zenith, but always becoming a rich purple when
reaching there, and always curling round, leaving a
clear strip of sky, which may be described as four
fingers held at arm's length. The northern side from the
zenith was also illuminated with beautiful colors,
always curling round at the zenith, but were considered
to be merely a reproduction of the southern display, as
all colours south and north always corresponded. It was
a sight never to be forgotten, and was considered at the
time to be the greatest aurora recorded. These displays
were somewhat frequent sixty years ago, but altogether
inferior in every way to the aurora of '59. The one in
Perth caused some consternation to some people, and
the one in '59 caused feelings much the same to some.
The rationalist and pantheist saw nature in her most
exquisite robes, recognising the divine immanence,
immutable law, cause, and effect. The superstitious and
the fanatical had dire forebodings and thought it a
foreshadowing of Armageddon and final dissolution. –
Yours, etc.,*

C. F. Herbert
No. 1 Dangan-street.

On 21st November 1861 the remarkable actions of this aurora event were published in a paper and submitted to the Royal Society. It was further concluded that one of the factors that made this aurora event possible was that almost a week earlier another powerful aurora episode had taken place, which had 'cleared the path' for this second, more powerful storm. It must have been truly astonishing to witness such a historic event. Especially one so prepossessing. But all is not lost, for if you wish to experience such an event yourself, the aurora records kept by many governments around the globe provide some potentially exciting news. These records state that given the frequency and intensity of historical aurora activity, it can be expected that an aurora display on par with that of 1859 may transpire on average every 150 years, with a reasonable range of occurrence of 100-250 years overall. Given the time this very sentence is being written, we are now roughly 160 years into this range, meaning an aurora display as brilliant as the great geomagnetic storm of 1859 could be due any moment. So much so, that on 23rd July 2012, a gigantic CME blasted out of the Sun and slipped past Earth without a single aurora being produced. The only reason we know this CME existed is because the STEREO-A spacecraft happened to be idly floating around at the time, before being unexpectedly bombarded by the extreme force of the CME, which it later reported to Earth. Upon receiving and analysing this report from the STEREO-A spacecraft, Daniel Baker of the University of Colorado stated that 'In my view the July 2012 storm was in all respects at least as strong as the 1859 Carrington event… the only difference is, it missed'. This solar storm further led another physicist, Pete Riley, to publish a paper in *Space Weather* titled, *On the Probability of Occurrence of Extreme Space Weather Events*. The research undertook looked at the probability of an enormous solar storm on par with that of 1859 occurring within the next ten years. The answer, which is rather exciting but also moderately concerning: 12%. However, if the CME of July 2012 had been ejected around a week earlier it would have collided with our planet instigating both mass havoc and splendid auroras, simultaneously. No doubt, if the CME had struck us, it would have been the most beautiful natural disaster of our lives, as a 2013 economic study by the USA estimated the total damage caused would have been roughly $US2.6 trillion. A further study by the Chinese State Key Laboratory of Space Weather projected that the damage caused would have been so significant, it would have taken between four and ten years to repair. So, hold on to your hat because it could be on its way right now. The thing is, there wouldn't be much notice if a gigantic solar storm along the lines of 1859 was to hit Earth. This is because magnetic field strength and the orientation of incoming plasma are two key ingredients of our aurora forecasting tools. This doesn't mean we would not know an aurora storm was on its way, it just means we could only truly gauge its intensity around 15-30 minutes before impact. So, leave yourself a good jacket close to the door, just in case. Because tonight could be one of the most spectacular nights in human history.

The above information poses a question: *what would the repercussions be if such an event occurred today? What would happen to our technology dependent world? Would we still be able to operate as we currently do without interruption?* In short, the answer is no. Firstly there would be a combination of high-energy sunlight, consisting of mostly x-rays and ultraviolet light, that would interfere with the Earth's radio communications systems. Shortly after, the first people to suffer the ill effects of such a strong solar flare would be the astronauts who are currently floating around in the International Space Station – as well as any other astronauts that happened to be in space at the time. These astronauts would be highly susceptible to the pending radiation storm as they would not be fully protected by Earth's magnetic field. It would then be the general population's turn to suffer the ill effects, as the super massive coronal mass ejection would collide with Earth and cause powerful electromagnetic fluctuations. The reason this would be so detrimental in today's society is because we practically live inside a cyber cocoon. GPS systems would falter and fail causing distress to mobile phone users, satellite navigation systems in cars, and more importantly, aircraft tracking systems. Satellite communications would stop working, causing disruption to many businesses around the globe, and the electric grid would drop out of use leaving billions without power and heat. Your internet would disappear, your freezer would defrost, your lights would go out, most of us wouldn't be able to cook food, and more or less everything around you would cease to function as our world is almost wholly dependent on electricity. Basically, the world would start to fall to pieces. But at least there would be some fancy lights in the sky to take the edge off things.

BEYOND EARTH

'Two possibilities exist: either we are alone in the Universe or we are not. Both are equally terrifying.'

Arthur C. Clarke

You may be surprised to hear that Earth is not the only planet we know of which regularly experiences aurora activity. Astronomers over the years have witnessed and photographed auroras on Jupiter, Saturn, Uranus, Neptune, Venus, Mars, and Io (one of Jupiter's currently known 62 Moons) using the Hubble Space Telescope. By doing this the scientists have discovered that the auroras can essentially appear on any planet in the Universe, as long as it has somewhat of an atmosphere. But there is also the other consideration that some form of energy must be able to exert influence over the planet, which is obviously not the case for every planet out there. But given the Milky Way alone is believed to possess around 100-200 billion planets, and the observable Universe is thought to hold around 10^{24} planets, there is a fairly high chance there is still a lot of aurora activity going on. And all that is required is that the planet or moon's atmosphere is bombarded with energetic particles. If the planet has a magnetic field (such as Earth), then the auroras will hit and be controlled by that field, giving them a similarly distinct shape to the auroras seen on our planet. If the planet has no magnetic field, then the auroras will appear as an uncontrolled glow, taking any shape or form they wish. If we are to look at the history of our planet and the origins of our magnetic field, we will discover our field has only been present for approximately four billion years. This means any aurora shows that occurred more than four billion years ago would not have been as controlled and shapely as they are today, with the beautiful contours and streamlined ribbon profiles we have spoken about in this book. They would have taken the form of the aforementioned, uncontrolled glow, making areas of the skies of Earth radiate like a gigantic light bulb.

It's also interesting to learn that the auroras of other planets can appear in different colours to those of Earth, as the atmospheres of other planets can contain different particles and atoms. Furthermore, the unique colour a gas produces upon collision or reaction is known as its 'spectrum'. This means that different gases and combinations of gases will release diverse spectrums of colour upon collision. It's by no means naive to say that the auroras of the Universe could appear in an infinite number of colours, or shades of colours, if there are an unlimited number of gas combinations to be found. And considering the observable Universe is thought to be 13.7 billion light years across, and currently expanding at the speed of light, this has the potential to be an exceedingly real possibility.

Finally, on a closing note, the strongest aurora performance ever detected belongs to a nearby brown dwarf star with the catchy name, LSR J1835+3259. The auroras of this star are the first extrasolar auroras ever monitored in human history. They were uncovered in July 2015 by the Karl G. Jansky Very Large Array radio astronomy observatory (another catchy name) in New Mexico, USA. The method used to monitor these distant auroras was the analysing of emitted radio waves, and unbelievably, it was found that the auroras of this brown dwarf were red in colour and were estimated to have been one million times brighter than any aurora ever observed on Earth.

It's incredible to think that over the past 400 years or so, since the auroras have been assessed scientifically, we have come so far and gained so much knowledge. And as technology improves our progress will only continue. That being said, we now understand the auroras to a certain degree, but there are still questions we have not yet answered. There's a certain amount of joy contained within these remaining questions, however, as they have the ability to cause our imaginations to flourish – just like the cavemen, Arctic tribes, Indians, Vikings, and Aborigines' minds had done before ours, as they sought answers about the dancing lights they witnessed in the sky. Their ascribing of the auroras to dragons, gods, ancient spirits, daring swans, dwarfs, and dancing fish was their way of deciphering the mystery of the lights. Their way of trying to make sense of a world they were born into without request. But what we do know today is that solar flares, solar wind, and solar cycles contribute to aurora activity. We also know that electrons, oxygen and nitrogen atoms, and Earth's magnetic field, make the auroras possible, and that they can be seen predominantly over the Polar Regions of Earth as well as elsewhere in the Universe. We know there's been much work, ingenious thought, and creativity poured into the process of trying to understand the lights, and that some of the characters we have encountered along the way have been the greatest scientists, philosophers, and thinkers the world has ever known. And knowing these things is astonishing. Our progress as a species has been dumbfounding. If it were left to me to discover the source and operation of the auroras I would not know where to begin. But, after researching and writing this book there is one thing I could not be surer of: regardless of all the knowledge, science, and facts we have about the auroras, I feel the more we learn, the more amazing but less mysterious the lights become. I have realised that the day we have the knowledge to answer every question asked will be an exceptionally sad day indeed. Don't get me wrong, it's a privilege to be able to access this information and be living in an age with profound amounts of data, whilst having the world's knowledge at our fingertips thanks to modern technology. But the child inside me doesn't care for facts and figures. The child inside me isn't concerned with progress and explanations. If the truth be told, the child inside wishes the auroras were not formed by the science we have discussed, but by a giant, happy, football playing walrus.

PHOTOGRAPHS

The great landmass of Antarctica, close to where Captain Cook and William Wales first documented the aurora australis in the Southern Ocean. These documentations went down in history as the first 'official' observations of the southern lights in western society.

A time-lapse photograph of the aurora borealis. This photo showcases the multiple colours the auroras can display. The layer of the atmosphere solar wind interacts with is a strong contributor to the colour the auroras appear. So, varying colours mean different aurora beams are being displayed at contrasting heights.

A desolate Arctic tundra acting as a backdrop to a stunning aurora display.

This photograph was taken during the middle stages of an aurora substorm. This can be seen as the aurora has fully uncurled and is stretching like an archer's bow towards the eastern and western horizons.

The entwined aurora above tells us this is an early stage of the aurora substorm. Soon the lights will untwine into a straighter line stretching towards the eastern and western horizons, before continually folding upon themselves and then diffusing into the night.

Planet Earth. Our home. Sitting within a region of the Sun's influence means we are barraged by its sometimes-aggressive activity. This activity can then interact with the magnetic poles of our planet and cause the auroras to appear in the skies.

Iceland was one of the lands the Vikings called home, which made it one of their ancient aurora viewing platforms. Thanks to this, Iceland became one of the birthplaces of aurora mythology, research, and discovery.

It is believed that the Milky Way is home to 100-200 billion planets and 400 billion stars, many of which experience the auroras just like we do.

A spectacularly clear night sky displaying the auroras in Greenland. Many of the Arctic tribes of Greenland believed the auroras were caused by the dancing of departed souls, and whilst they were displayed, the tribes would sit in complete silence as a sign of respect to the dead.

A lone spectator witnessing an aurora display.

A series of aurora beams dancing above a pair of remote lodges in Norway.

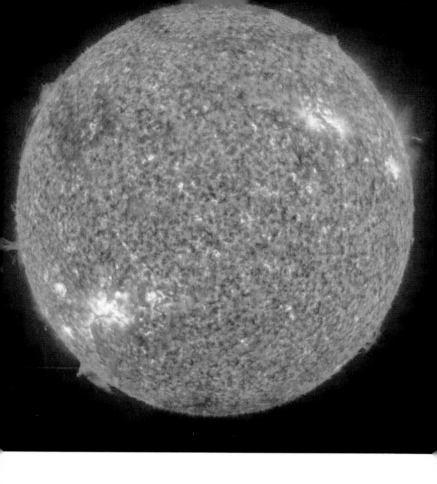

The Sun is the root cause of the auroras. Without its solar flares and coronal mass ejections there would be no auroras here on Earth.

An image of the Sun littered with sunspots. If you look closely, you can see the magnetic field lines that permeate the sunspots. These sunspots will increase in number over an 11-year period causing greater and greater solar activity, and therefore, stronger auroras on Earth.

ACKNOWLEDGEMENTS

I would like to pay special thanks to my beautiful wife, Roxana, for her unwavering support and belief throughout the entire research, writing and publishing process. As a debut author, the whole experience of taking a book from conception to publishing proved to be more difficult and longer winded than I ever imagined at the offset. Without your support, I am not sure it would have been possible.

I would also like to give a special thanks to my mum, dad, and brother, Sean, for everything you have done over the years. Without you, I would not possess the life, sarcasm and patience to complete such a work.

Lastly, I would like to thank my readers. If it were not for you then this would be nothing more than a stagnant text, waiting to be discovered. It takes a reader to bring a book to life; and I hope you have enjoyed it. If you have, please feel free to review my work and tell your friends and family. As a debut self-publishing author, your small contribution is worth more than you could ever imagine.

Thank you.

Now go and find the auroras.

CONNECT WITH THE AUTHOR

Liamdevonport.com

Liam Devonport

SELECT BIBLIOGRAPHY

Abruptearthchanges.files.wordpress.com. (2019). [online] Available at: https://abruptearthchanges.files.wordpress.com/2018/05/tchijevsky-physical-factors-of-the-historical-process.pdf

Aristotle. (2004). *Meteorology*. [Belle Fourche]: NuVision Publications.

Aurora Service (Europe). (2019). *Northern Lights Folklore - Aurora Service (Europe)*. [online]

Baltimore American and Commercial Advertiser (1859). The Aurora Borealis. p.2.

Birkeland, K. and Muir, J. (1908). *The Norwegian aurora polaris expedition, 1902-1903*. Christiania: H. Aschehoug & Co.

Boole, G. (1952). *The laws of thought, 1854*. La Salle, Ill.: Open Court Pub.

Choi, C. (2019). *First Alien Auroras Found, Are 1 Million Times Brighter Than Any On Earth*. [online]

European Space Agency. (2019). *Mars Express discovers aurorae on Mars*. [online]

Feldstein, Y. (1986). A quarter of a century with the auroral oval. *Eos, Transactions American Geophysical Union*, 67(40), p.761.

Feldstein, Y. and Galperin, Y. (1999). Comment on `Magnetospheric source region of discrete auroras inferred from their relationship with isotropy boundaries of energetic particles' by A. G. Yahnin. *Annales Geophysicae*, 17(1), p.37.

Gilbert, K. (2019). *The Layers of the Atmosphere*. [online] WorldAtlas.

Green, J. and Boardsen, S. (2006). Duration and extent of the great auroral storm of 1859. *Advances in Space Research*, 38(2), pp.130-135.

Green, J., Boardsen, S., Odenwald, S., Humble, J. and Pazamickas, K. (2006). Eyewitness reports of the great auroral storm of 1859. *Advances in Space Research*, 38(2), pp.145-154.

Green, L. (2017). *15 million degrees*. 1st ed. London: Penguin Books Ltd.

Hall, C., Pederson, D. and Bryson, G. (2001). *Northern lights*. 1st ed. Seattle: Sasquatch Books.

Hayakawa, H., Ebihara, Y., Hand, D., Hayakawa, S., Kumar, S., Mukherjee, S. and Veenadhari, B. (2018). Low-latitude Aurorae during the Extreme Space Weather Events in 1859. *The Astrophysical Journal*, 869(1), p.57.

Hayakawa, H., Iwahashi, K., Tamazawa, H., Isobe, H., Kataoka, R., Ebihara, Y., Miyahara, H., Kawamura, A. and Shibata, K. (2016). East Asian observations of low-latitude aurora during the Carrington magnetic storm. *Publications of the Astronomical Society of Japan*, 68(6), p.99.

IFLScience. (2019). *A Solar Storm Put A Crack In Earth's Magnetic Field*. [online]

Jago, L. (2013). *The northern lights*. 1st ed. New York: Vintage Books.

Luminarium.org. (2019). *Aurora Borealis, the Northern Lights, in Mythology and Folklore.*. [online]

Moreno Cárdenas, F., Cristancho Sánchez, S. and Vargas Domínguez, S. (2016). The grand aurorae borealis seen in

Colombia in 1859. *Advances in Space Research*, 57(1), pp.257-267.

NASA. (2019). *What is a coronal mass ejection or CME?*. [online]

News.nationalgeographic.com. (2019). *Auroras Make Weird Noises, and Now We Know Why*. [online]

News.nationalgeographic.com. (2019). *What If the Biggest Solar Storm on Record Happened Today?*. [online] Available at: https://news.nationalgeographic.com/news/2011/03/110302-solar-flares-sun-storms-earth-danger-carrington-event-science/ [Accessed 24 Mar. 2019].

Northernlightsforecast.com. (2019). *The Myths*. [online]

Notablebiographies.com. (2019). *Anders Celsius Biography - life, name, death, son, book, information, born, time, year*. [online]

Odenwald, S. and Green, J. (2008). Bracing for a solor superstorm. *Scientific American*, 299(2), pp.80-87.

Oschman, J. (2016). *Energy medicine*. Edinburgh: Elsevier.

Science.nasa.gov. (2019). *Near Miss: The Solar Superstorm of July 2012 | Science Mission Directorate*. [online]

Severe space weather events. (2009). Washington, D.C.: National Academies Press.

Siscoe, G. (1978). An historical footnote on the origin of "Aurora Borealis". *Eos, Transactions American Geophysical Union*, 59(12), p.994.

Spaceplace.nasa.gov. (2019). *What is an aurora? | NASA Space Place – NASA Science for Kids*. [online].

Stamper, R., Lockwood, M., Wild, M. and Clark, T. (1999). Solar causes of the long-term increase in geomagnetic activity. *Journal of Geophysical Research: Space Physics*, 104(A12), pp.28325-28342.

Stamper, R., Lockwood, M., Wild, M. and Clark, T. (1999). Solar causes of the long-term increase in geomagnetic activity. *Journal of Geophysical Research: Space Physics*, 104(A12), pp.28325-28342.

Störmer, C. (1946). Frequency of 12,330 measured heights of aurora from southern Norway in the years 1911–1944. *Journal of Geophysical Research*, 51(4), p.501.

The Daily News (1909). The Great Aurora of 1859. p.9.

User, S. (2019). *Mythology of the Northern Lights*. [online] Theaurorazone.com.

Windridge, M. (2017). *Aurora*. 1st ed. London: Harperpress.

Liam Devonport

Made in United States
Cleveland, OH
12 December 2024

11423992R00079